Biodiversity in India

Volume 2

T. Pullaiah

2024
Regency Publications
A Division of
Astral International Pvt. Ltd.
New Delhi – 110 002

First Published, 2003
Reprinted, 2024

ISBN: 9788187498711 (HB)

Publisher's Note:

Every possible effort has been made to ensure that the information contained in this book is accurate at the time of going to press, and the publisher and author cannot accept responsibility for any errors or omissions, however caused. No responsibility for loss or damage occasioned to any person acting, or refraining from action, as a result of the material in this publication can be accepted by the editor, the publisher or the author. The Publisher is not associated with any product or vendor mentioned in the book. The contents of this work are intended to further general scientific research, understanding and discussion only. Readers should consult with a specialist where appropriate.

Every effort has been made to trace the owners of copyright material used in this book, if any. The author and the publisher will be grateful for any omission brought to their notice for acknowledgement in the future editions of the book.

Published by : **Regency Publications**
A Division of
Astral International Pvt. Ltd.
– ISO 9001:2015 Certified Company
4736/23, Ansari Road, Darya Ganj
New Delhi-110 002
Ph. 011-43549197, 23278134
E-mail: info@astralint.com
Website: www.astralint.com

PREFACE

I am happy to note that the first volume on Biodiversity in India got tremendous response from the readers. Encouraged by this we are bringing out this second volume. The present volume includes 5 articles covering species diversity, genetic diversity and ecosystem diversity. The first article by Mary and Vijaya Koteswari deals with the genetic variability and biosystematics of *Dactyloctenium aegyptium*. Vasudeva has given phytogeographical and cytopalynological characteristics of Pteridopohytes in Satpura hills. The fourth article by Subba Rao *et al.*, gives the genetic diversity of Pearl millet and its wild relatives. The last article is on Aquatic flowering diversity in coastal plains of Konkan, Goa and Karnataka by Gaikwad *et al.* I am thankful to all the authors for their contributions. I invite authors to contribute to the future volumes on Biodiversity in India.

14th April, 2003 **T. Pullaiah**

CONTENTS

LIST OF CONTRIBUTORS

T.N. Mary
Department of Botany
Nagarjuna University
Nagarjuna Nagar-522510
Andhra Pradesh

M. Vijaya Koteswari
Department of Botany
Hindu College
Guntur-522006

S.M. Vasudeva
Department of Botany
Punjabi University
Patiala-147002
Punjab

M.V. Subba Rao
Department of Botany
Andhra University
Visakhapatnam-530003

P.V. Arjuna Rao
Department of Botany
Andhra University
Visakhapatnam-530003

V. Manga
Department of Botany
Andhra University
Visakhapatnam-530003

S.P. Gaikwad
Department of Botany
Shivaji University
Kolhapur-416004

M.M. Sardesai
Department of Botany
Shivaji University
Kolhapur-416004

S.R. Yadav
Department of Botany
Delhi University
Delhi-110007

CHAPTER 1

THE STUDIES ON GENETIC VARIABILITY IN CROWFOOT GRASS (*DACTYLOCTENIUM AEGYPTIUM* (L.) BEAUV.) (POACEAE) FROM GUNTUR, ANDHRA PRADESH

T.N. Mary and M. Vijaya Koteswari***

Introduction

No two individuals in sexually reproducing populations are the same. The underlying factor in this diversity is genetic, though other factors play their role in divergence of populations. This biodiversity is the outcome of natural evolution which has been going on for the last 3.5 billion years. Charles Darwin's theory of evolution by natural selection (1859) is predicted upon the existence of genetic diversity, a major component of Biodiversity.

'Biological diversity' means the variability among living organisms from all sources, including inter alia, terrestrial, marine and other aquatic ecosystems and the ecological complexes of which they are part; this includes diversity within species, between species and of ecosystems.

Biodiversity convention make efforts to conserve biodiversity on sustainable basis and to attribute value to the biological resources of the developing countries. The Government of India is taking necessary steps for biodiversity conservation. In order to prevent depletion of biodiversity, it is necessary to understand how variation in natural populations is maintained. All the species we observe on the planet

* Department of Botany, Nagarjuna University, Nagarjuna Nagar-522510, A.P.
** Department of Botany, Hindu College, Guntur-522006

today are the result of the processes of evolution through time and modern taxonomic procedures seek to establish the patterns of evolutionary relationship between them.

From the present studies on crowfoot grass (*Dactyloctenium aegyptium* L. Beauv.) the prevailing genetic variability is analysed and appraised through the observations on natural populations growing in and around Guntur area.

It is obvious from the existing literature (see Sharma & Salam, 1984; Sharma, 1984) that the crowfoot grass, *D. aegyptium*, an annual, growing in the plains of Punjab State in India, actually represents a species-complex, comprising *D. aegyptium*, its closely allied congener, *D. aristatum*, and some taxa with morphological characters intermediate between these two species. This species-complex is still an unresolved problem. The various chromosome number reports available in literature for this species-complex, such as n = 10, 18, 19, 20, 22, 23, etc., in all probability, do not belong to *D. aegyptium* alone, but to a few other intermingled taxa, as also opined by Sharma & Salam (1984). Biosystematic survey of *D. aegyptium* complex by the above workers has revealed that it is not possible to separate the taxa contained in this complex on the basis of the identification key given by Bor (1960).

According to Bor (1960) *D. aegyptium* is stoloniferous with spikes 2–5 cm long and tip of rachis extending up to 2 mm in length. On the other hand, *D. aristatum* is non-stoloniferous with spikes 0.5–2.0 cm long and tip of rachis produced up to 4 mm. However, the separation of the two species on the basis of above characters has been found difficult.

What actually prompted the present study is:

(1) As mentioned already, there are difficulties involved in distinguishing the various taxa within *D. aegyptium* complex, (2) There is practically no information about *D. aegyptium* from South India, and (3) *D. aegyptium* was found to show staggering complexity among its local interbreeding populations, inhabiting Nagarjuna University campus area of 292.7 acres (A.P. State, South-east India). It is, therefore, felt worthwhile investigating the nature and causes of variability noticed in *D. aegyptium* complex, and critically to evaluate the identities of the taxa within this complex.

The present investigation with a general ecological slant was initiated with the following objectives:

1) To study the morphological variability of *D. aegyptium* complex and to assess whether the differences among these populations are significant or not.
2) To determine whether the morphological differences among these populations persist when grown in uniform environment and

also at reciprocal sites, i.e. to determine whether the variation is genetically fixed or environment-induced.

3) To study the chromosomal number, their meiotic behaviour and fertility relationships among these forms.
4) To assess the ecological behaviour of the species through the study of its seed germination, reproductive and aggressive capacities, anatomy, phenology and phytosociology, etc., and
5) To evaluate the possible causes of morphological divergence in this grass complex.

Description of the Site of Investigation

Location

The site of investigation is the Nagarjuna University campus. It is 17 km away from Guntur town (lat. 16°18′ N long. 80°28′ E) on the Madras-Howrah National Highway (NH-5) in Andhra Pradesh State, South India. The campus has an area of 292.7 acres (188.5 ha) with approximate bearings of lat. 16°24′ and long. 80°33′ E (Fig. 1 and

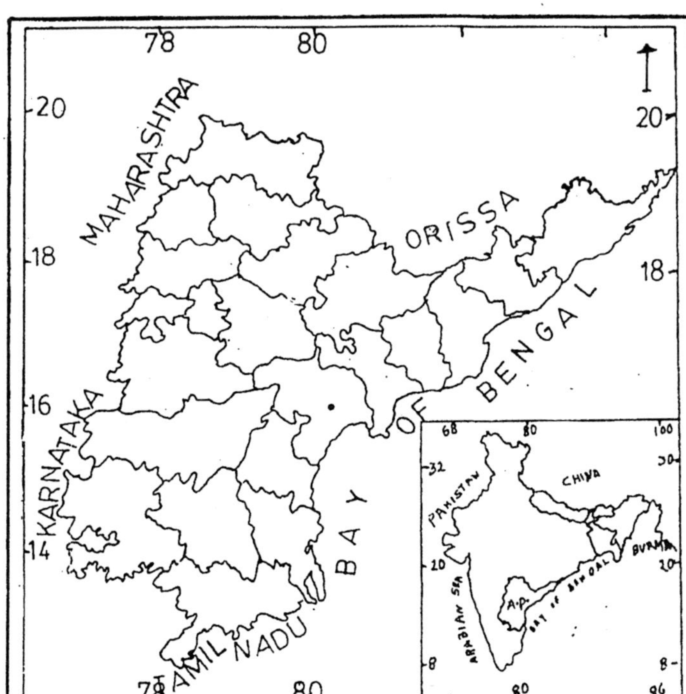

Fig. 1: Location of the study site; Inset India.

Fig. 2) at an elevation of 34.54 m above mean sea level. The area, in general has even topography with gentle gradient from west to east and south-west to north-east.

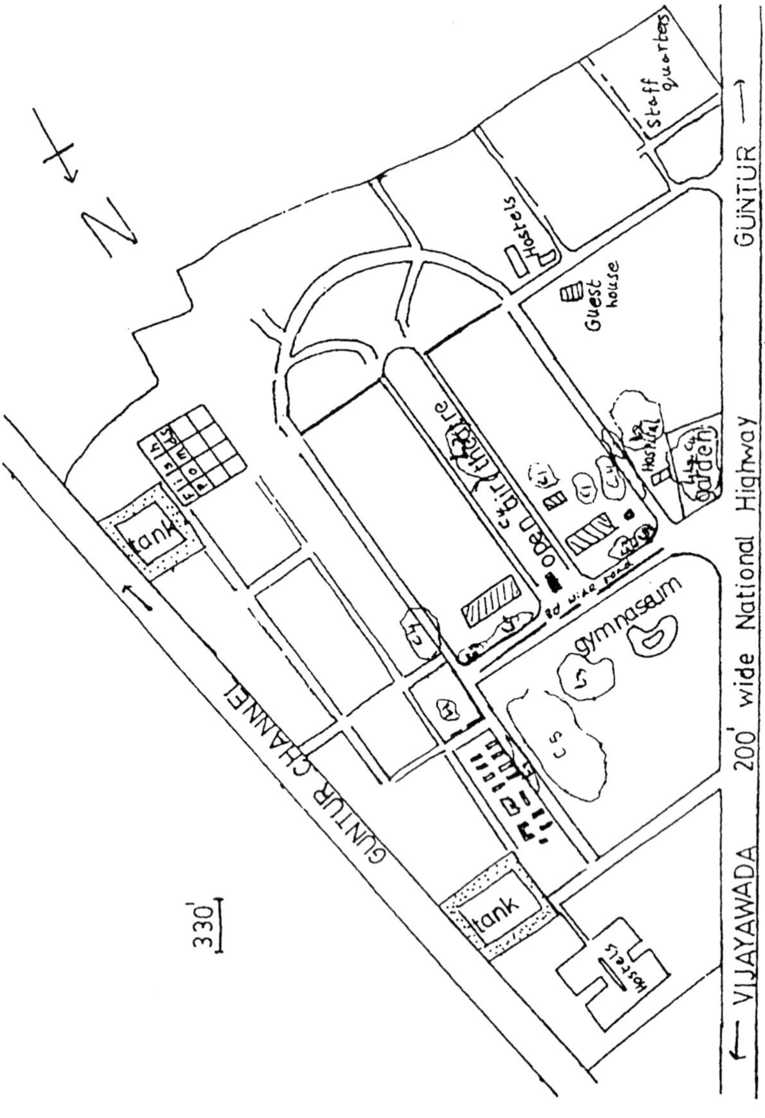

Fig. 2: Layout of Nagarjuna University and the sites of occurrence of different forms of *D. aegyptium* complex.

Geological Features

The soils are red lateritic. The campus area is boardered on the northern side with black soils where cotton and vegetables are grown, and on the east is an irrigation canal from River Krishna; the area beyond this canal is under rice cultivation. The southern side is marked by a vast expanse of uncultivable land with an undergrowth of grasses and a few weeds, but essentially dominated by the palmtrees (*Borassus flabellifer*). Black granite and, to a little extent, calcareous nodules occur in the vicinity of the campus area. The soils are generally alkaline with pH ranging from 7.5 to 8.5.

Meteorological Features

The area experiences three seasons, viz. rainy season, winter and summer. Rainy season generally starts in the month of June and ends in October. It is characterised by relatively hot and humid climate with minimum fluctuations in diurnal temperature and humidity. Vegetation appears and reaches its prime growth during this season. Winter, extending from November to February, is moderately cool and dry, except for occasional heavy rains in November owing to cyclones. Most of the annuals enter into reproductive phase during winter. Summer begins in March and continues till the onset of monsoon in June. It is the hottest and driest part of the year with high rate of evapotranspiration and longer photoperiods. Most of the herbaceous species wither away and die, leaving very little green to cover parched soil. Overall, it is a drought-prone area.

The meteorological data collected from the nearest meteorological observatory (CTRI, Guntur) during he study period are given in Table 1 and a graphic representation of the same in Fig. 3.

General Vegetation

This grassland area has been subject to serious disturbance and interference by way of mechanical clearing and burning of weeds, and construction activity. These conditions favour secondary succession. Its general features of the flora are indicated by the genera that are essentially xerophytic and drought-affected. The important tree species are *Acacia nilotica* (L.) Willd., *A. horrida* (L.P.) Willd., *Albizia lebbeck* (L.) Willd, *Borassus flabellifer* L., *Phoenix sylvestris* (L.) Roxb. and the South african naturalized species *Prosopis juliflora* (SW) DC. the climbers and shrubs are sources of valuable medicinal weeds (Rao, 1981a, b). Major medicinal weed species include *Withania somnifera* (L.)

Table 1: Meteorological data for the years 1981, 1982, 1983 and 1984

Month	1981					1982					1983					1984				
	Temp. Max. (°C)	Temp. Max. (°C)	RH (%) at 08.30	RH (%) at 17.30	Rain-fall (mm)	Temp. Max. (°C)	Temp. Max. (°C)	RH (%) at 08.30	RH (%) at 17.30	Rain-fall (mm)	Temp. Max. (°C)	Temp. Max. (°C)	RH (%) at 08.30	RH (%) at 17.30	Rain-fall (mm)	Temp. Max. (°C)	Temp. Max. (°C)	RH (%) at 08.30	RH (%) at 17.30	Rain-fall (mm)
January	29.82	18.18	86.51	67.77	14.20	30.18	17.87	91.45	61.52	—	30.30	18.20	83.00	49.00	—	29.89	19.60	89.78	53.39	—
February	33.66	18.16	89.32	68.79	—	32.60	20.38	90.82	56.46	—	32.69	21.15	87.28	46.14	—	31.10	20.22	87.51	52.24	7.80
March	36.40	22.15	90.75	69.90	4.00	34.36	23.47	74.07	52.03	—	36.46	24.48	87.49	49.48	—	35.58	22.90	87.52	45.46	2.00
April	39.29	25.39	90.13	59.47	20.00	38.20	26.22	83.70	51.33	—	37.85	26.85	88.00	47.00	—	37.83	26.13	84.30	45.46	2.00
May	45.96	27.54	83.78	52.30	3.70	39.35	27.38	82.52	45.74	82.72	42.10	29.10	80.00	45.60	5.20	41.30	28.30	84.80	41.60	2.00
June	41.19	29.35	79.43	57.87	83.20	38.32	28.36	83.00	56.80	70.08	38.70	28.03	80.63	61.26	106.50	40.50	29.80	66.60	57.60	12.60
July	38.23	26.49	87.55	62.29	76.50	35.74	26.18	80.03	59.60	149.80	35.56	26.26	83.30	63.30	198.50	34.90	25.07	82.70	53.20	259.40
August	36.45	25.98	84.68	67.52	120.40	34.66	26.21	77.55	61.88	63.00	34.08	26.32	85.58	67.94	129.40	35.70	26.25	78.61	54.29	68.60
September	34.66	26.28	84.00	72.53	253.00	33.76	25.35	85.43	67.53	190.00	32.78	26.38	88.90	67.73	347.20	34.26	25.96	83.16	56.96	189.40
October	13.04	24.07	87.87	72.23	155.90	32.50	23.90	88.00	62.00	267.90	32.08	29.85	87.23	70.13	161.30	32.60	23.38	88.00	56.00	64.50
November	0.98	20.57	86.10	74.28	—	31.00	22.00	89.00	65.01	86.30	31.20	21.96	82.76	64.00	24.60	29.91	21.05	86.83	58.06	14.20
December	9.92	19.19	86.77	61.13	—	30.20	19.00	87.00	44.00	—	29.25	19.80	83.70	57.67	11.40	31.03	19.24	88.48	52.00	—

Temp. = Temperature, Max. = Maximum, Min. = Minimum, RH = Relative Humidity

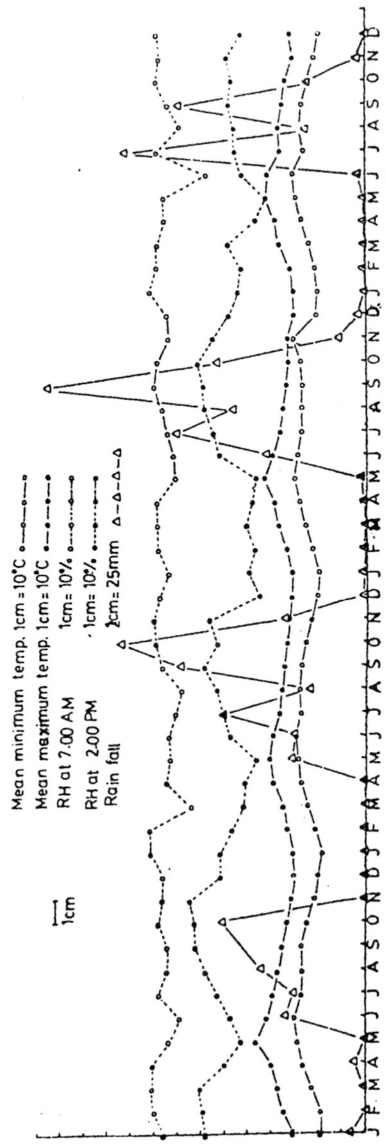

Fig. 3: Graphical representation of meteorological data for the years 1981–1984.

Dun., *Hybanthus ennaespermus* (L.) F.W. Muell., *Aristolochia indica* Linn., *Hemidesmus indicus* (L.) R.Br., *Tylophora asthmatica* W & A etc.

The above accounts is to impress upon the pedestrian nature of the vegetation of the study area, mainly clothed with the grasslands interspersed with a wide variety of weeds (Plate 1a).

Systematic Position, General Morphology, Geographic Distribution and Economic Importance of *Dactyloctenium aegyptium* and *D. aristatum*

Systematic Position

The genus *Dactyloctenium* Willd. belongs to the family Poaceae and is represented by 10 species, which are well distributed in tropics (Willis, 1966). Hooker (1897) assigned the genus to the tribe Chlorideae due to spikelets having one to many flowers while Bor (1960) included in Eragrosteae, considering the digitate nature of spikes. In India, this genus is represented by five valid species (Bor, 1960), viz., *D. aegyptium* (L.) Beauv., *D. australe* Steud., *D. henrardianum* Bor, *D. sindicum* Boiss., and *D. aristatum* Link. Among these, *D. aegytpium* is very common. Important synonyms of this species are cited below:

Dactyloctenium aegyptium (L) Beauv., Ess. Agrost. Expl. pl. 15 (1812).
Cynosurus aegyptius Linn. Sp. pl. ed. 1, 72 (1753).
Eleusine aegyptia (Linn.) Desf., Fl. Atlant, 1, 85 (1798).
Eleusine pectinata Moench, Meth. pl. Suppl. 68 (1802).
Chloris mucronata Michx., Fl. Bor. Amer. 1, 59 (1803).
Dactyloctenium aegyptiacum Willd., Enum. pl. Hort. Beroll. 1029 (1809).
Dactyloctenium mucronatum Willd., Loc. cit. 1029.

General Morphological Description of the Plant

The crowfoot grass, *D. aegyptium*, is a spreading to slightly ascending annual grass, recorded all over India (Plate 1). *D. aristatum* is closely related to *D. aegyptium*, and restricted to North-west India (Bor, 1960) but found common to South India (Mary, *et al.*, 1996). *D. aegyptium* inhabits cultivable fields, canal banks, roadsides and desert areas. Culms compressed and spreading with ascending ends, shooting at the nodes, branching profuse and commonly forming radiating mats, often 20 to 40 cm high, sometimes even up to 1 m, sheaths compressed, smooth. Ligule as short membrane with jogged edge, ending in hairs, leaf blades flat, smooth or sparsely covered with soft, basally swollen hairs on both surfaces, but particularly dense on the margins and nerves beneath. Inflorescence digitate, coarse spikes 2 to 7, rarely 1, arranged in a finger-like fashion, spikes 1 to 3 cm long, 6 to 8 mm wide, spikelets sessile 3–5-flowered, laterally compressed, arranged in 2 rows at right angles to the dorsally flattened rachis; florets 2.5 to 3.0 mm long, glumes 1-nerved, first and second glumes long, empty, somewhat rough on the keel. Rachis mucronate at tip, the lower glume

Plate 1: Bird's eye view of the study site: a—Open grassland with *Cymbopogon* and *Dactyloctenium* grass, outlined by trees of *Cassia* species; b—Close view of crowfoot grass populations in nature; c & d—Disturbance in the grass land due to biotic influence.

Plate 2: a—*D. aristatum*; b & c—Variants.

Plate 3: The maintenance of short rachilla in compact group: a—C_1; b—C_2; c—C_3 forms.

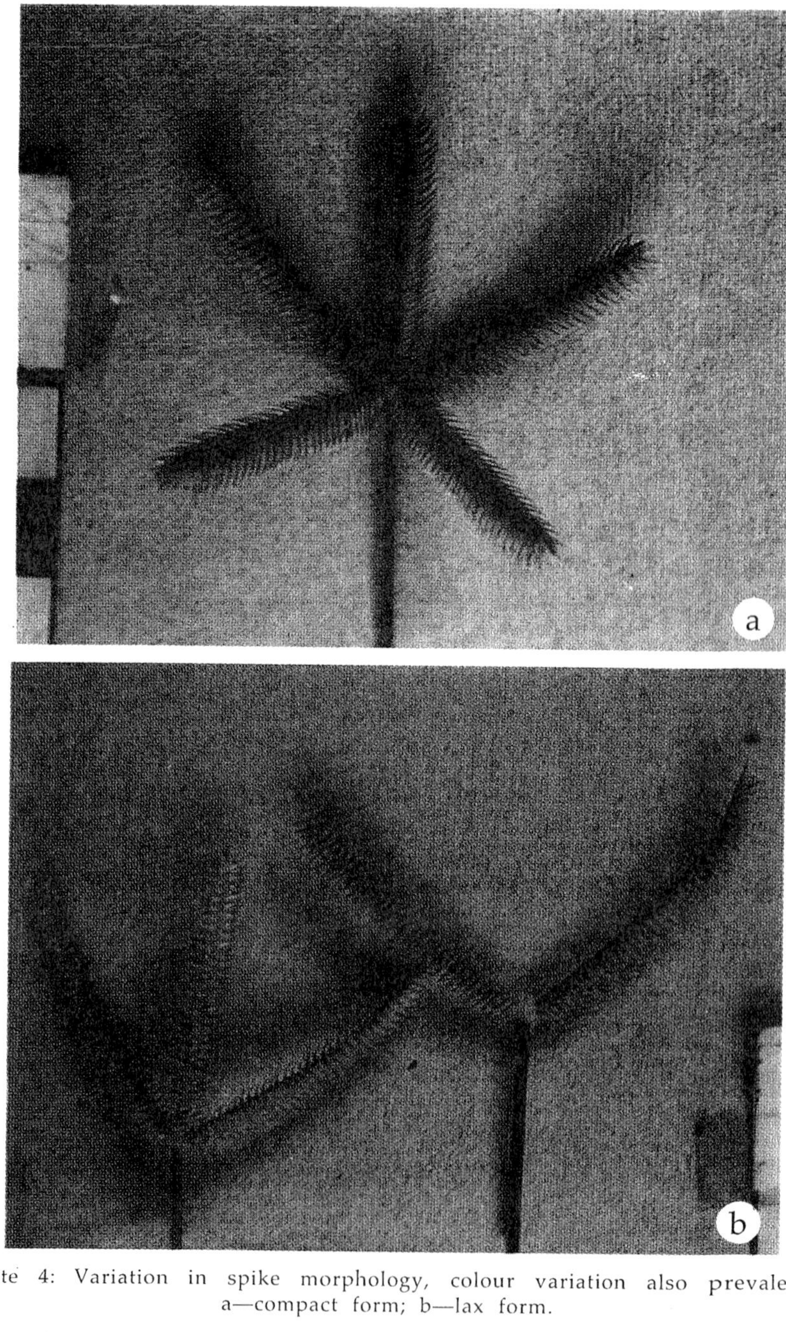

Plate 4: Variation in spike morphology, colour variation also prevalent: a—compact form; b—lax form.

Plate 5: Morphology of C1, C2, C3: a—Habit of C1; b—Habit of C2; c—Habit of C3.

Plate 6: Morphology C4 & C5: a—Habit of C4; b—Habit of C5.

Plate 6c, d: C5 showing stoloniferous growth and prostrate nature.

Plate 7: Morphology of L1, L2 and L3: a—Habit of L1; b—Habit of L2; c—Habit of L3.

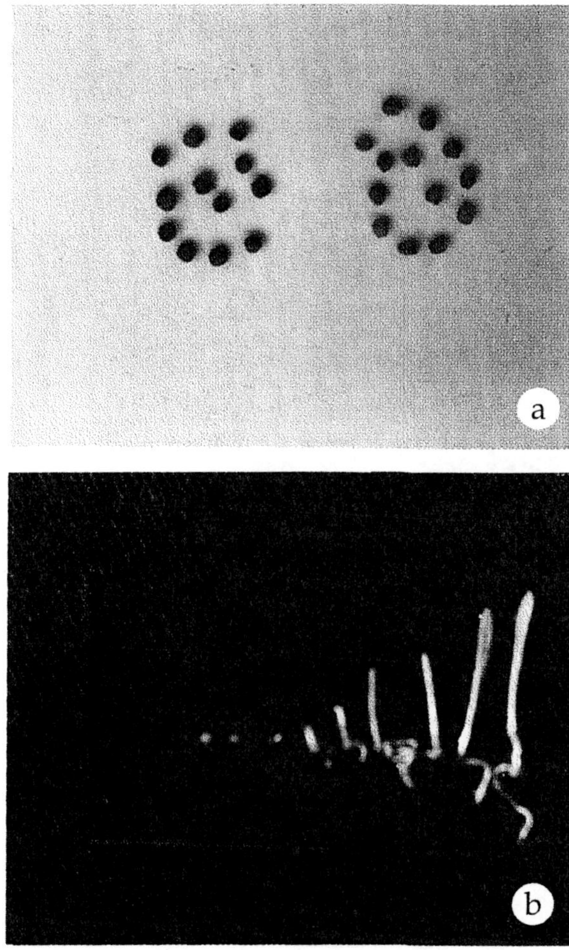

Plate 8: Seed colour variation in *D. aegyptium* complex: a—Orange brown: Compact forms; Light grey: Lax forms; b—Epigeal seed germination.

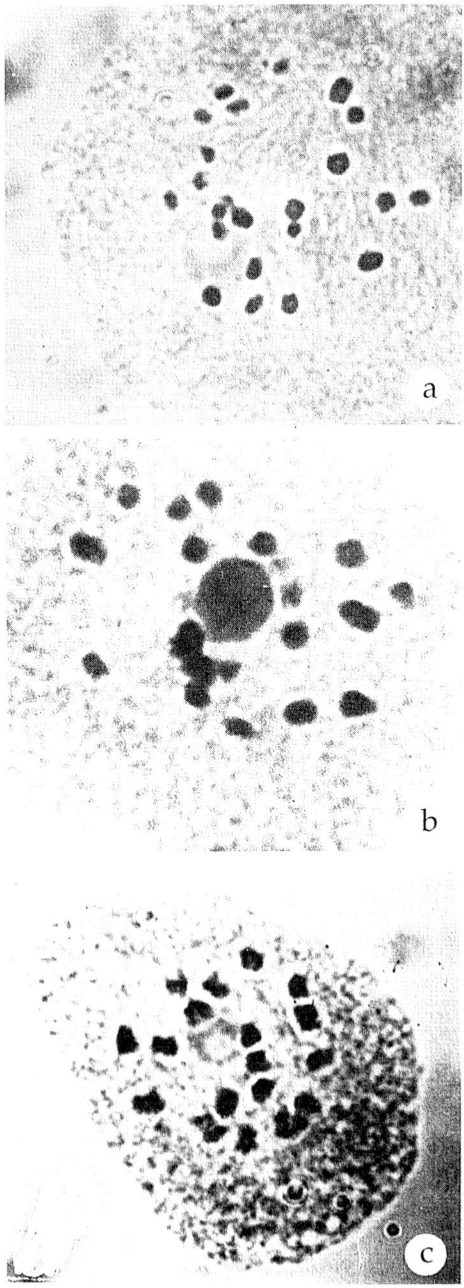

Plate 9: Pollen mother cells of C4 and L2 and L3 at diakinesis: a—C4 with 23 bivalents (×1000); b—L2 with 20 bivalents (×1500); c—L3 with 19 bivalents (×1200).

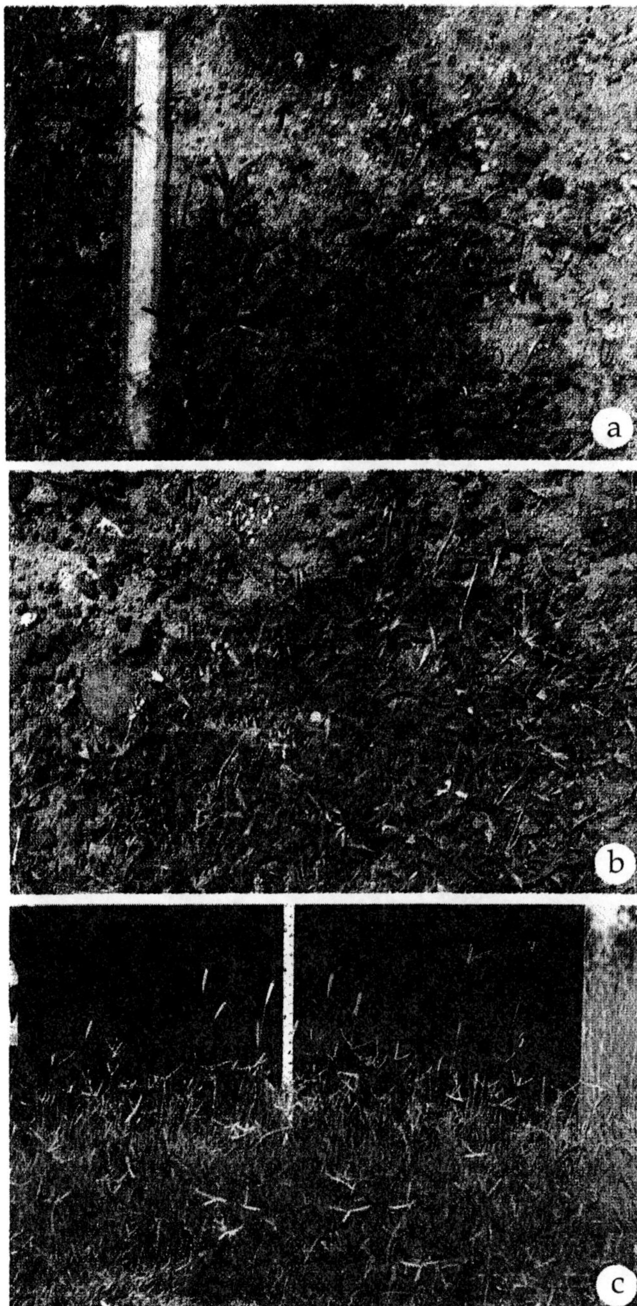

Plate 10: Seed dispersal & myrmecophily: a & b—; c—species of *Chloris* & *Cymbopogon* as associates.

Plate 11: C4(a), L2(b) and C5(c) in common garden conditions, maintaining their respective characters.

1.5 mm long and the upper 2.5 mm long with an equally long rigid awn. Lemma 3 mm long with a stiff sharp point or tip, often recurved, normally persistent. Palea hyaline, shorter than lemma, winged on keels. Lodicules 2, stamens 3, styles 2 and free. Stigmas plumose. Grain (caryopsis) about 1 mm long, obovate, loose pericarp, irregular, transversely ridged, orange brown.

The species can be readily recognized by its coarse, digitate spikes.

Geographic Distribution

It is an annual grass, native of Old World tropics. The principal range of distribution as a weed is from latitude 15°N to 15°S. Forty-five countries reported it as a common weed associated with 19 world crops (Holm *et al.*, 1977). The geographic distribution of *D. aegyptium* in various crops is shown in Fig. 4.

Economic Importance

It stands second in producing a palatable nutritious fodder to cattle (Sen, 1981), an is considered suitable for producing milk and for fattening (Banerjee, 1972). The seed is used in India and Africa as a food grain at times of want, but is said to have an unpleasant taste and cause internal disorders (Bor, 1960). It has medicinal importance in that it is used to cure ailments of kidney and ulcers (Holm *et al.*, 1977). The seed, if stored for a long period, proves toxic to human beings and cattle owing to the presence of cyanogenetic glycosides (Bor, 1960).

D. aristatum

D. aristatum Link., slender annual grass, culms erect, 4–10 cm, simple, nodes glabrous leaf sheath 0.5–1.75 cm long, ligule membranous, very short with fringe of hairs, blades linear-lanceolate, flat, 1.5–5.5, 0.1–5.5 cm, ciliate. Inflorescence a short terminal spike, 1.25–2 cm long, rachis slender, bearded at base, produced at apex into 2–4 mm long, arista beyond the spike. Grain globose with rugose margins, 1 mm light grey colour.

Review of Literature

Previous Work on the Crowfoot Grass, D. aegyptium

A perusal of the existing literature reveals that the information on ecosystematics is scanty vis-à-vis the information on diverse aspects of

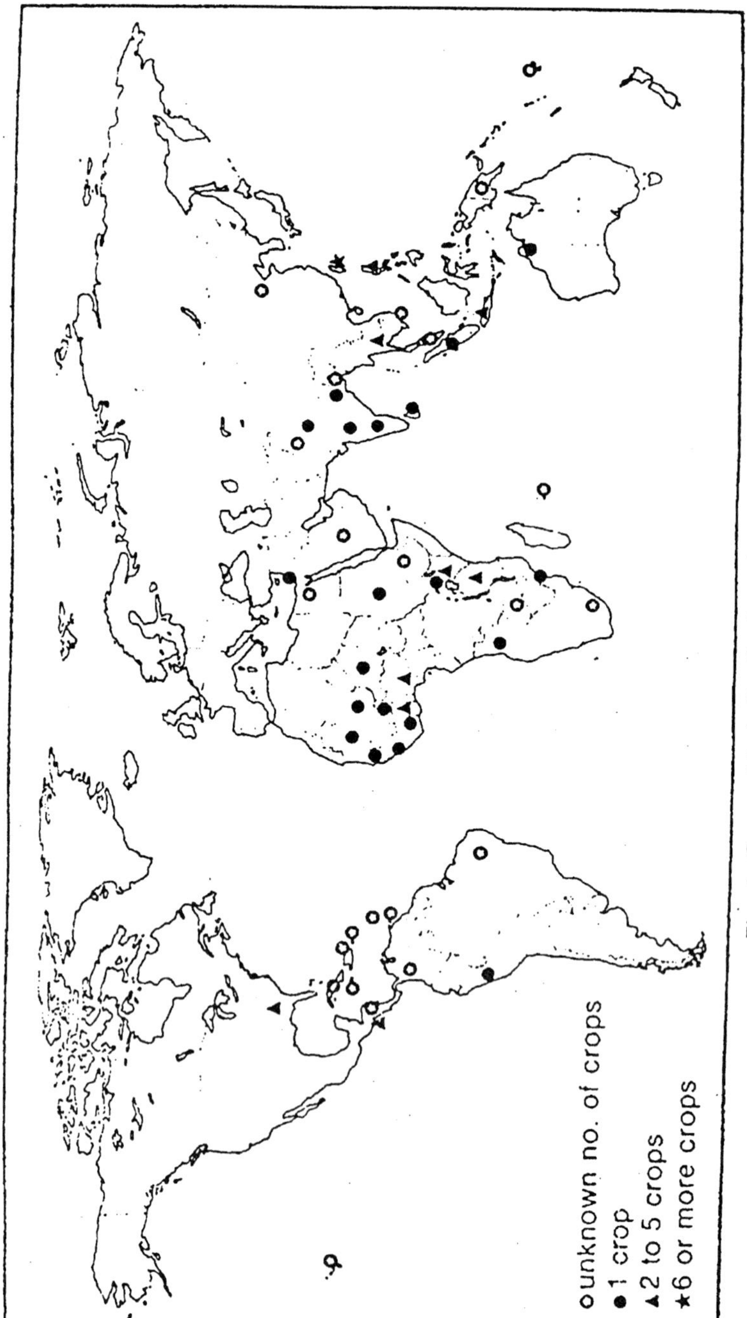

Fig. 4: Geographical distribution of *D. aegyptium.*

biology of this grass. Earlier works of Bor (1960), Holm *et al.* (1977), and Sen (1981) summarized the data on morphology and ecology of this species from North India.

The first anatomical report of the leaf in *D. aegyptium* was that by Grob (1896). Gunzel (1912–1913) also contributed to the knowledge of the leaf anatomy in the genus *Dactyloctenium*. Prat (1932, 1934, 1936) mentioned that the leaf resembles that of *Chloris* and *Cynodon*. The study on leaf epidermis in *D. aegytpium* was made by Fisher & Schweiekerdt (1941). Metcalfe (1960) made a significant contribution to the anatomy of this species, Sachdeva & Ritakals (1981) provided information on stem and leaf of three cytotypes in *D. aegyptium* complex from South Eastern part of Punjab.

This species has been pondered as a prolific seed producer, and seeds are regarded to remain viable for period ranging from 1–19 years (Holm *et al.*, 1977; Sen, 1981). Myrmecochory was implicated in seed dispersal (see Sen, 1981). Seed germination studies were made by Kumar *et al.* (1971); Holm *et al.* (1977); Singhal & Sen (1981), and Sant (1962, 1963, 1964 a, b, c, 1965). Nelvigi (1962) worked out the reproductive capacity of *D. aegyptium* at Varanasi and observed the maximum value in rainy season. Choudhri & Varshney (1986) studied the effect of soil salinity on growth and reproductive behaviour of this species. The phenology of this species was investigated from fields of Jodhpur by Sen (1981) and from Kurukshetra by Singh & Yadava (1974).

Many chromosome number reports on this grass are available from literature. The first report of 2n = 34 was by Krishnaswamy (1931). The other euploid and aneuploid reports are: 2n = 20 by Bor (1960), Gould (1960), 2n = 36 by Moffett & Hurcombe (1949), Chandola (1959), Bor (1960), Gould (1966), Mehra & Sharma (1975), 2n = 40 by Tateoka (1965a) 2n = 48 by Avudulov (1928), Parodi (1946), Bor (1960), Larsen (1963), Tateoka (1965a), Christopher & Abraham (1974), 2n = 52 by Sharma *et al.* (1978) and 2n = 38 by Olarode (1974), 2n = 18, 20, 38, 46 by Bir & Sahni (1987), 2n = 40, 46, 54 by Sachdeva & Ritakals (1981) under *D. aegyptium* complex.

A total of 105 populations of *D. aegyptium* and *D. aristatum* were studied for spike length and prolongation of rachis by Sharma (1984) and suggested the modification of Bor's key for separation of *D. aristatum* from *D. aegyptium*. Cytotaxonomic studies in 81 populations of *D. aegyptium* complex from various localities of Eastern Punjab have been carried out by Sharma and Salam (1984). They have reported four cytotypes (2n = 38, 40, 44, 46) in *D. aegyptium* and two (2n = 20, 40) in *D. aristatum*. Cyto-morphological and chemotaxonomic studies of three cytotypes in *D. aegyptium* complex from western part of Punjab were contributed by Sachdeva & Ritakals (1981). From other part of Punjab two cytotypes with 2n = 22, 46 were reported (Bir *et al.*, 1987) that showed no morphological distinction except for microcharacters. Thus the work on the crowfoot grass from South Eastern part of India is scanty.

From Guntur three cytotypes with 2n = 38, 40 and 46 chromosome numbers having significant morphological variation among them were reported (Vijaya Koteswari, 1990 & Mary *et al.*, 1992)

Material and Methods

Material of Study

The materials for the present study is the crowfoot grass, *Dactyloctenium aegyptium, D. aristatum* and other populations (Fig. 5 and Plates 2 & 3). In the present study eight morphological forms, differentiated can be grouped into two broad categories based mainly on the nature of the spikes, viz., compact and lax. Other characters such as growth habit, number of spikes, tillers, length of spike and rachilla, etc., have been taken into consideration in distinguishing the above forms. The compact forms are designated as C1, C2, C3, C4 and C5 and the lax forms as L1, L2 and L3. The highly diagnostic detailed features of each form are given under results.

Methods of Study

The methods adopted here are dealt with under four heads: analytical, statistical, cytological and autecological.

Analytical Methods

Composite soil samples were collected during 1983–84 from eight different native sites of the grassland area, which supported the different phenotypic forms of *D. aegyptium*. The method of collection, and physical and chemical analyses of the soils are same as given in Chopra & Kanwar (1976). Moisture content, water holding capacity and pH of the soils were determined on the fresh samples of soils whereas the analysis of mineral nutrients and organic matter was done on air-dried soils stored in the laboratory in air-tight bottles. Texture of the soil and the classification thereof was determined, following Chopra & Kanwar (1976), by passing the dry, crumpled soil through a set of sieves. Formulae used in the calculation of moisture content and water holding capacity are:

$$\% \text{ Moisture content} = \frac{\text{Loss of wt. of fresh soil sample on drying}}{\text{Dry wt. of the soil}} \times 100$$

$$= \frac{fwt - dwt}{dwt} \times 100$$

$$\% \text{ Water holding capacity} = \frac{\text{Wt. of soil samples after the excess water is drained} - \text{dwt of the soil}}{\text{dwt of the soil}} \times 100$$

$$= \frac{swt - dwt}{dwt} \times 100$$

Where
fwt = fresh weight of the sample
swt = saturation weight of the sample
dwt = dry weight of the sample

pH of the soil sample was determined with the help of an Elico pH meter (model L1–10). For weighing, a simple balance, a single-pan electrical balance, and a Boringor's balance were used. Available nitrogen was estimated by using alkaline permanganate method. Potassium was estimated by using flame photometric method. Olsen's method was employed for estimating the available phosphorus. For calcium and magnesium, EDTA method was used. Walklay and Black's tritration method was used for the estimation of organic matter.

The procedures of plant analysis for estimation of total nitrogen and phosphorus are after Jackson (1967) and Piper (1942).

Statistical Methods

Random samples were used in all the experiments. Procedures for determining mean, standard error were those of Schefler (1969) and Kershaw (1973).

Values given after ± in the tables are those of the standard error. 'F' test was used to compare the samples and to test the significance of difference.

Cytological Methods

For meiotic studies, young spikes were fixed in 1:3 acetic alcohol. After 24 hours they were transferred into 70% alcohol and smeared in

Fig. 5: Habit *Dactyloctenium aegyptium* (L.) Beauv.

2% acetocarmine solution. Attempts were also made to study mitotic behaviour. Microphotographs were taken from fresh preparations.

Autecological Methods

For general autecological studies, standard methods given in Costing & Reed (1952), Misra (1968) and Pandeya *et al.*, (1968) were followed. Sources of certain other methods are from Harborne (1973), Suryanarayana & Krishnaswamy (1948) and Murthy & Mujumdar (1962) and others.

Morphology

Thirty plants were picked up at random from each population of *D. aegyptium* at reproductive stage for an appraisal of a variety of quantitative morphological characters. The following structural parameters were chosen for comparison, e.g., length of root, shoot, flag leaf, peduncle, spike, rachilla, number of tillers, leaves, spikelets etc.

The mean and SE values of the quantitative data for 30 individuals of each form were calculated. 'F' values were also calculated to assess the level of significance between these forms, and some of the characters were projected through histograms.

Common Garden Experiment

To find out whether the different forms of *D. aegyptium* would maintain their original characters, the seeds and genets of these forms were grown in July 1982 in the University garden under uniform environmental conditions. The seedlings were transplanted at random in plots and also in pots, and were allowed to grow. At the end of December 1982, all survivors were harvested and analysed for the morphological characters.

Reciprocal Transplant Experiments with Tillers

In July 1982, sixty clones of each of the forms of C4, C5 and L2 were removed from the field sites and numbered. Identical numbers were given to the tillers of the same form. Standardization was further increased by growing them in greenhouse conditions for one month. They were watered daily. The plants were replanted at the field sites according to the plan shown below:

Sites of origin	C4	C5	L2
Sampling of plant material	|	|	|
Standardization of material	60	60	60
Transplanting tillers	|	|	|
Transplant sites	C4 C5 C2 20 20 20	C4 C5 L2 20 20 20	C4 C5 L2 20 20 20

Some soil was transferred with the plants to minimise root damage. It was not possible to interpolate transplants without disturbing the surrounding vegetation. Therefore, a standard zone, 2–4 cm in width, was cleared around each individual transplant. Outside this zone, vegetation was undisturbed. Transplants were watered once and were grown for six months. No death of the clones were recorded since they were transplanted during rainy season. Each single experiment involved 60 tillers and they were planted in the sites of origin and the other sites in a randomized block design. Twenty planting positions arranged in a rectangle with four rows and five columns (within and between row distance 10 cm) constituted a replicate. One-third of a single genotype was assigned to the home site, and the remainder were equally distributed for transplanting between two away-sites as indicated in the plan given above. This ensured that genetic variation among the plants from the same origin (population) was exactly the same at the different transplant sites.

The survival of these transplants was recorded at one month intervals. In the month of December 1982, all the surviving transplants were harvested and the morphological characters such as lengths and numbers of roots, tillers, leaves, spikes and flowers were recorded at the time of harvest.

The idea given by Grime (1979) was followed in studying the stress response in *D. aegyptium*. The growth performance in particular, and the seed production at high, medium and low densities were compared.

Based on the data so collected the relative growth rate was determined as per the formula shown below:

$$RGR = \frac{\log e\ W_2 - \log e\ W_1}{t_2 - t_1}\ g/day$$

Where, RGR = Regressive growth rate; t = time; w = dry weight

Relative growth rate indicates the increase in dry weight per unit time.

To study the effect of organic matter on the growth performance, pots were filled with sand, and FYM, and (red soil) separately. The seeds were sown and in each pot two seedlings were allowed to grow. Then the various parameters were studied as in the case of density studies.

The ratio of leaf weight to plant weight (LWR) compares the dry weight of leaves with the dry weight accumulation of the entire plant. Similarly the spike/leaf weight ratio compares the dry weight of the spike with that of leaves.

Estimation of Nitrogen and Phosphorus in Plant Parts

Freshly collected plants were cleaned and dried in hot air oven at 80°C. Dry plant samples were analysed for nitrogen by micro-kjeldahl method (Jackson, 1967) and for phosphorus by the method given in Fogg and Wilkinson (1958).

Anatomy

Simple free hand sections in the required plane were taken for anatomical study of root, stem and leaf. Since this grass species offers a great difficulty in making epidermal peelings to study the stomatal features, the special method as given by Suryanarayana & Krishnaswamy (1948) was adopted. Temporary preparations stained in safranin and mounted in glycerine were observed under a compound microscope and the illustrations of anatomical features were drawn with the aid of camera lucida.

Pollen Studies

The pollen grains were collected at 10.30 A.M. from freshly dehisced anthers of two forms of *D. aegyptium* viz., C4 and L2, typifying compact and lax groups respectively. Pollen size refers to the pollen diameter in µm. Pollen fertility was scored by using 2% acetocarmine. Half-stained and unstained grains were considered as non-viable.

Seed Studies

Seed Output

The average number of seeds per spike was determined and it was multiplied by the number of spikes and number of flowering tillers.

The Seed Diameter and Weight

The seed diameter was determined with the help of standardized ocular micrometer; 25 seeds were measured for each form and the average value was taken into consideration. The seeds were weighed on an electrical monopan balance, taking three hundred seeds for each form in three sets, and the average value was calculated.

Germination Studies

Germination experiments were conducted in petridishes in between two moist filter papers. Three replicates of 50 seeds each were maintained for each experiment and the observations were recorded daily until there was no germination. Protrusion of the radical was taken as the basis of germination. It was studied under continuous darkness by wrapping the petridishes in black paper and also under different wavelengths, using red, green, blue and yellow cellophane papers. Continuous light was supplied by a 40 W white fluorescent bulb which was kept 0.5 m away from the petridish. Different light intensities were obtained by keeping the petridishes in three different places. The seeds were scarified mechanically with sand in a mortar for about 5 minutes.

To study the effect of water stress, the seeds were kept on filter papers of varied thickness and the same quantity of water was added. To know the effect of salinity, seeds were soaked in different concentrations (0.05 to 0.2%) of sodium chloride. The effect of IAA, and Kn on seed germination was studied with presoaked seeds for 24 hours in different concentrations of 100, 200, 300, 400 and 500 ppm. For all the sets, a control was maintained.

Reproductive Capacity

Following the formula given by Salisbury (1942), the reproductive capacity was calculated as the product of average seed output and the fraction represented by the average percentage of germination. Here the germination percentage obtained for the seeds stored for 16 weeks was considered. The formula employed is as follows:

$$\text{Reproductive capacity} = \frac{\text{Average seed output} \times \text{\% germination}}{100} \times 100$$

Aggressive Capacity

Salisbury (1942) defined aggressive capacity as the potentiality of a species to colonize and is represented by the product of its average reproductive capacity and average percentage survival of its seedlings in the natural habitats. As soon as the seedlings emerged after the showers, 10 permanent quadrats of 50 cm × 50 cm size were laid near the parent plants. In each quadrat, a fairly large number of seedlings were noted at the beginning and countings were made at successive intervals. After six weeks the percentage survival of seedlings was calculated. The aggressive capacity was then calculated by employing the formula given below:

$$\text{Aggressive capacity} = \frac{\text{Average reproductive capacity} \times \text{Average percentage survival of seedlings}}{100}$$

Chlorophyll Estimation

The chlorophyll content of the leaves, stems and the inflorescence of the plants of approximately, 30, 60 and 90 days old were studied from natural populations of the different forms. One gram of the material of the prescribed stage was made into a thin paste in 100 ml of 80% acetone in a mortar. It was filtered through a Watman filter paper No. 1 in a Bruckner's funnel and the colour intensity of the filtrate was measured at 645 nm and 663 nm wave lengths on a spectrophotometre. The values in each case were the means of three replicates. The chlorophyll *a* and *b* were calculated following Harborne (1973) as given below:

$$\text{Chlorophyll } a \text{ (mg/g)} = \frac{12.7 \times \text{O.D. at 663 nm} - 0.86 \times \text{O.D. at 645 nm}}{\text{a} \times 1000 \times \text{w}} \times \text{V}$$

$$\text{Chlorophyll } b \text{ (mg/g)} = \frac{19.3 \times \text{O.D. at 645 nm} - 3.6 \times \text{O.D. at 663 nm}}{\text{a} \times 1000 \times \text{w}} \times \text{V}$$

Where,
V = volume
W = weight of the plant material
a = length of the light path, and
O.D. = optical density

Here the sum of the chlorophyll *a* and *b* is considered as the total chlorophyll content.

Phenology

The phenograms were constructed for a period of one year for all the forms separately. As the natural populations were generally subject to a variety of biotic influences, studies under cultural conditions were undertaken to evaluate phenological calendar along with the total life span of the different forms of *D. aegyptium*. The seeds were put in earthen pots for germination. From the data of sowing the important events and their respective dates of occurrence were noted until the plants' death.

Phytosociology

For phytosociological studies, the methods given by Braun-Blanquet (1932) were followed. A metal frame quadrat of 50 cm^2 was used to collect the data. The flora associated with *D. aegyptium* in the study area were listed. The nomenclature of the plants identified is in accordance with Gamble (1957) and Hooker (1897). The number of species and individuals, and their basal areas were recorded in the field. From this, frequency, relative frequency, density, relative density, abundance, relative dominance and Importance Value Index (IVI) were calculated according to the formulae given by Misra (1968). For each form 10 quadrats were sampled. The plants beyond seedling stage i.e. more than 2 cm height in case of monocots, and beyond first leaf stage in dicots were counted. Tiller analysis of Singh & Yadava (1974) was adopted in counting the number of grass species. Since it was not possible to trace the beginning and end of most of the dominant grasses like *Cymbopogon, Heteropogon, Iseilema* etc., the branches with individual functional root systems were counted as individuals. For the basal area, the plant's diameter at 2 cm height from the soil was considered. The associates were arranged in alphabetical order.

Sorenson (1948) gave a simple formula to establish the index of similarity between two communities as

$$S = \frac{2C}{A + B}$$

Where,
S = similarity index
A = number of species in first group
B = number of species in second group, and
C = number of species common to both groups

Results

Morphology

D. aegyptium is an annual, appearing generally in rainy and winter seasons. A critical study of its populations revealed the occurrence of eight morphologically distinct forms, recognized by their size, stature, colour, spike morphology, number of tillers, spikes/inflorescence, etc. The plants are basically divided into two main groups based upon the spike morphology-compact and lax types (Fig. 6 and Plate 4).

The compact forms, C1 to C5, are characterized by the presence of compactly arranged spikelets on rachis (Fig. 6a). The salient features of the compact forms are as given below:

C1: Dwarf, non-stoloniferous, erect, slender with 4–5 leaves and only one compact, short spike with 1 mm length of rachilla and with a few (5–8) spikelets, dark green in colour and occurring in dense patches under optimum moisture content (Plate 5d).

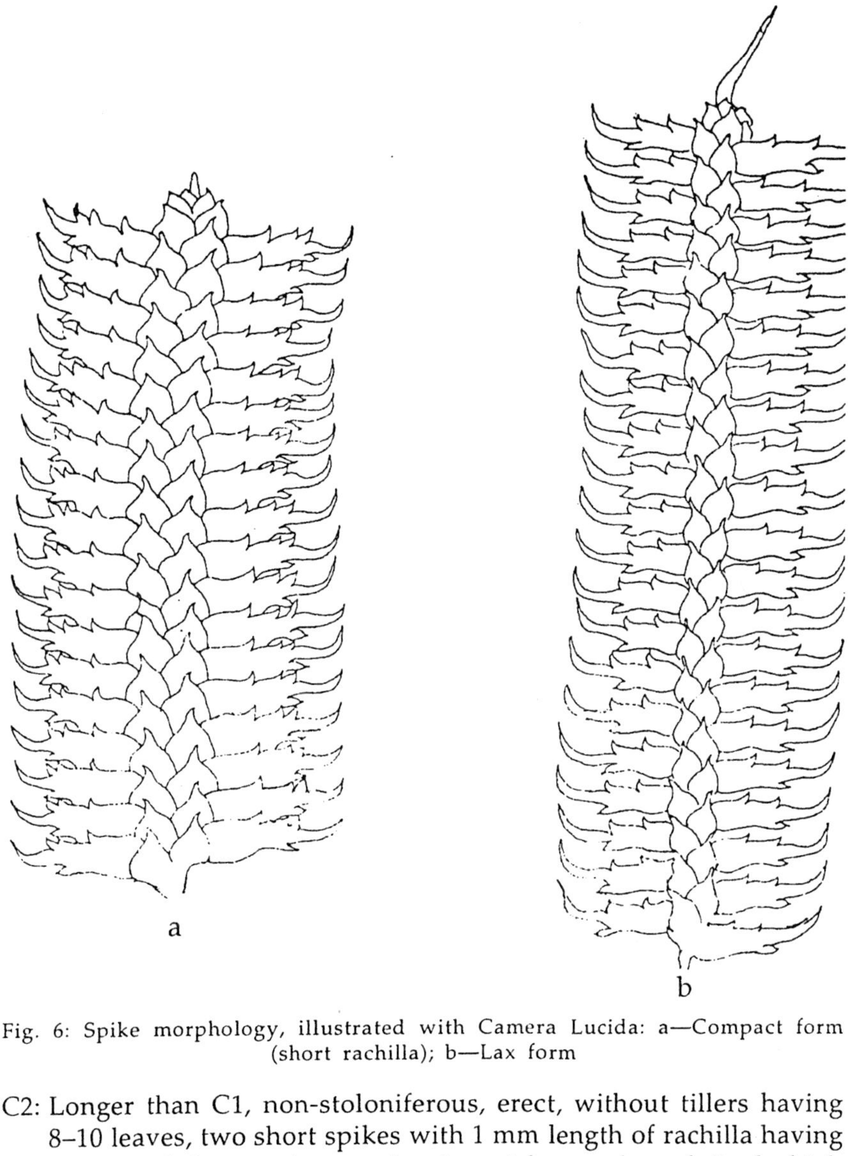

Fig. 6: Spike morphology, illustrated with Camera Lucida: a—Compact form (short rachilla); b—Lax form

C2: Longer than C1, non-stoloniferous, erect, without tillers having 8–10 leaves, two short spikes with 1 mm length of rachilla having 20–30 spikelets, and occurring in patches under relatively high moisture content (Plate 5b).

C3: Medium in height non-stoloniferous, having about 4 tillers, 3–4 short spikes, with 1 mm rachilla, and having 34–37 spikelets and occurring along roadsides (Plate 5c).

C4: Large in size, stoloniferous, erect, with 5–10 tillers, thick green leaves, 4–7 spikes having 1 mm rachilla, spikes and internodes longer than in other compact forms, having about 60 spikelets, occurring under high moisture content (Plate 6a).

C5: Stoloniferous, Prostrate, forming dense patches, 2–4 spikes having nearly 30 spikelets/spike and occurring in areas of low moisture content and high biotic interference (Plate 6b, 6c & 6d).

The lax forms, L1 to L3 are characterized by long spikes, loosely arranged spikelets on rachis with long rachilla (Fig. 6b).

L1: Non-stoloniferous, erect with 3–8 leaves, light green in colour, without tillers mostly single long spike varying from 1.0 cm to 4.0 cm length having 25–35 spikelets and rachilla upto 4 mm length (Plate 7a).

L2: Non-stoloniferous, erect, with 8–25 leaves, 2–5 tillers, and 1–3 long spikes, nearly 4 cm. with a long rachilla of 4 mm and 50 spikelets/spike (Plate 7b).

L3: Stoloniferous, prostrate, forming mats, with nearly ten tillers, 2–4 medium size spikes, 32 spikelets/spike and 2 mm length of rachilla, short internodes, greatly affected by trampling and grazing (Plate 7c).

Roots adventitious, stems (or culms) and leaves of compact forms dark green as contrasted with pale green in lax forms. Sometimes the forms with red pigmentation on culms, leaves, and spikes, were found in both the groups (Plates 1a, 1b & 4b).

Leaf sheath compressed and glabrous in both the groups. Hairiness of leaf varying in different forms, as shown in Table 2. Leaf hairs unicellular, distributed on the leaf margins as well as on the upper and lower surfaces in all the forms. However, in lax forms they are not only longer and denser at the leaf base, but profusely distributed through out the rest of the leaf area. In compact forms, leaf hairs relatively short and distributed sparsely on the upper surface.

Inflorescence having spikes arranged in digitate fashion (Fig. 6a). Spikelets secund in both the groups, and arranged compactly in compact group and loosely in lax group. The detailed description of the spikes is given below:

Table 2. Distribution of hairs on the leaf in the compact and lax forms of *D. aegyptium*

Position	Compact forms	Lax forms
Leaf lamina	Sparsely arranged; uniform in size, short	Densely arranged; different in size; long.
Margins	Throughout; short	Throughout; long
Midrib	Absent	Present

Compact Spikes

Spikelets flat, densely crowded on one side of the floral axis and spreading at right angles to the axis, thick green, 3–5-flowered (Fig. 7a). Rachilla about 2 mm long, i.e. just at the level of the spike. Mostly spikes of compact forms shorter than those of the lax forms; except C4; spikelets not overlapping at the centre of the spike (breadthwise) (Plate 4a).

Lax Spikes

Spikelets flat but not crowded as in compact forms, 2–3-flowered (Fig. 7b). Rachilla long, 4 mm or more in length, acute. Spike narrower

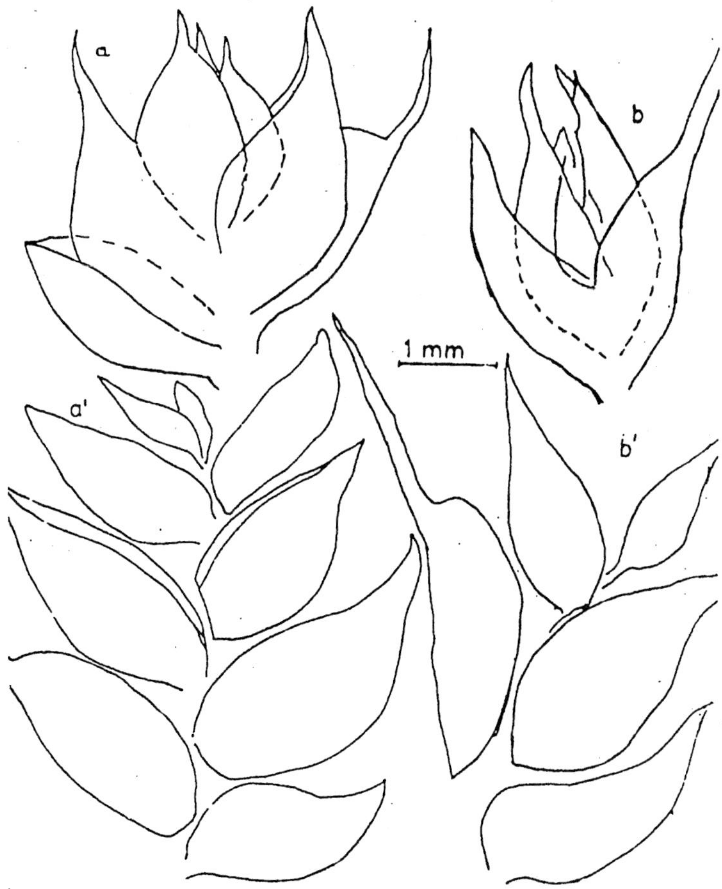

Fig. 7: Spikelet: a—Undissected spikelet of C4, a1—Dissected spikelet of the same; b—Undissectd spikelet of L2, b1—Dissected spikelet of the same

than in compact forms; spikelets overlapping at the centre of spike (Plate 4b).

Seeds (grains) rugose, subglobose, irregular in their outline. Seed colour varying in the two groups, being orange brown in compact forms and light gray in lax forms (Plate 8).

Morphological data collected for all the eight forms are summarized in Table 3. Based on important characters, comparative histograms and phytographs are constructed as given in Figs. 8 to 10. The above data bring out the quantitative comparison of the morphological features of the forms in question. 'F' test, when applied between the populations (Table 3), showed significant differences for root, shoot, leaf, spike, etc.

An examination of the tabulated data reveals that there are significant differences among the forms for the length of shoot, leaf, spike, internode and rachilla, number of flowers/spike, spikes/inflorescence, etc.

Culture Experiments

Common-garden Experiment

Under uniform environmental conditions, C4, C5 and L2 alone retained their respective characters (Plate 11). It can be said that environmental factors were less effective when the characters like the nature of spike(s) length of rachilla were considered, while some of the other characters like length of root, culm, spike, number of tillers and leaves appeared to be very much under environmental control.

Anatomy

The anatomical observations showed no differences in leaf and root except in the clum as shown in Fig. 11. The sclerenchymatous hypodermis was continuous in L2 whereas it was broken in C4 and C5. The ground tissue was larger in C5 comprising a layer of chlorenchymatous tissue while it was totally absent in C4 and L2. The size of the vascular bundle was larger in C4 when compared with other forms. The prostrate form showed reduced vascular bundles.

Cytology

Meiosis was studied from pollen mother cells (PMCs) in all the populations. 23 and 20 bivalents were regularly found in PMCs of compact and lax forms (Plate 9a & b). In addition, 19 bivalents were

Table 3. Mean standard error and F-values of certain characters in the eight populations of *D. aegyptium* complex

Population	Root length (cm)	Shoot length (cm)	Number of tillers	Number of leaves	Leaf length (cm)	Leaf breadth (cm)	Flag leaf length (cm)	Flag leaf breadth (cm)	Leaf lamina	Number of Inflorescences	Number of spikes	Length of spikes (cm)	Number of spikelets	Length of rachilla (mm)	Colour of seeds
1	5.17	7.17	1.00	6.07	2.01	0.22	0.79	0.11	Sparsely	1.00	1.00	0.56	8.10	0.10	Light
	± 0.23	± 0.31	± 0.00	± 0.19	± 0.11	± 0.11	± 0.07	± 0.03	hairy	± 0.00	± 0.00	± 0.53	± 0.20	± 0.00	brown
2	7.92	29.42	1.00	7.90	5.58	0.38	2.04	0.26	"	1.00	2.00	1.61	31.80	0.10	"
	± 0.34	± 0.97	± 0.00	± 0.31	± 0.38	± 0.01	± 0.18	± 0.01		± 0.00	± 0.00	± 0.06	± 1.42	± 0.00	
3	11.20	32.83	3.83	32.73	4.93	0.44	1.97	0.26	"	4.07	3.47	1.65	36.83	0.10	"
	± 0.41	± 0.74	± 0.23	± 1.81	± 0.21	± 0.01	± 0.13	± 0.01		± 0.35	± 0.09	± 0.06	± 1.26	± 0.00	
4	12.40	74.70	7.87	75.60	12.77	0.67	5.58	0.40	"	8.40	4.10	2.76	61.00	0.10	"
	± 0.51	± 2.60	± 0.62	± 0.64	± 0.67	± 0.03	± 0.30	± 0.01		± 0.69	± 0.06	± 0.04	± 1.53	± 0.00	
5	15.63	26.40	6.20	120.93	4.67	0.47	1.53	0.32	"	10.10	4.00	1.56	30.37	0.10	"
	± 0.60	± 1.06	± 0.28	± 6.01	± 0.11	± 0.02	± 0.09	± 0.01		± 1.04	± 0.00	± 0.07	± 0.90	± 0.00	
6	5.17	21.75	1.47	11.	4.19	0.36	2.80	2.27	Profusely	1.33	1.00	2.10	34.70	0.38	grey
	± 0.19	± 0.67	± 0.11	± 0.89	± 0.14	± 0.01	± 0.11	± 0.01	hairy	± 0.81	± 0.00	± 0.08	± 1.35	± 0.01	
7	9.20	32.23	2.73	24.40	6.41	0.48	3.91	2.37	"	3.70	2.27	3.38	51.63	0.40	"
	± 0.44	± 0.95	± 0.23	± 1.82	± 0.25	± 0.02	± 0.12	± 0.02		± 0.25	± 0.08	± 0.07	± 1.31	± 1.00	
8	8.15	12.15	10.23	113.33	3.08	0.42	2.25	0.35	"	22.77	2.00	1.87	32.23	0.20	"
	± 0.41	± 0.71	± 0.56	± 6.30	± 0.12	± 0.01	± 0.08	± 0.01		± 1.11	± 0.00	± 0.06	± 1.18	± 0.00	
F– value	72.55	311.19	106.65	127.36	118.48	60.49	97.41	21.26	—	145.31	200.11	59.86	169.33	2181.82	—

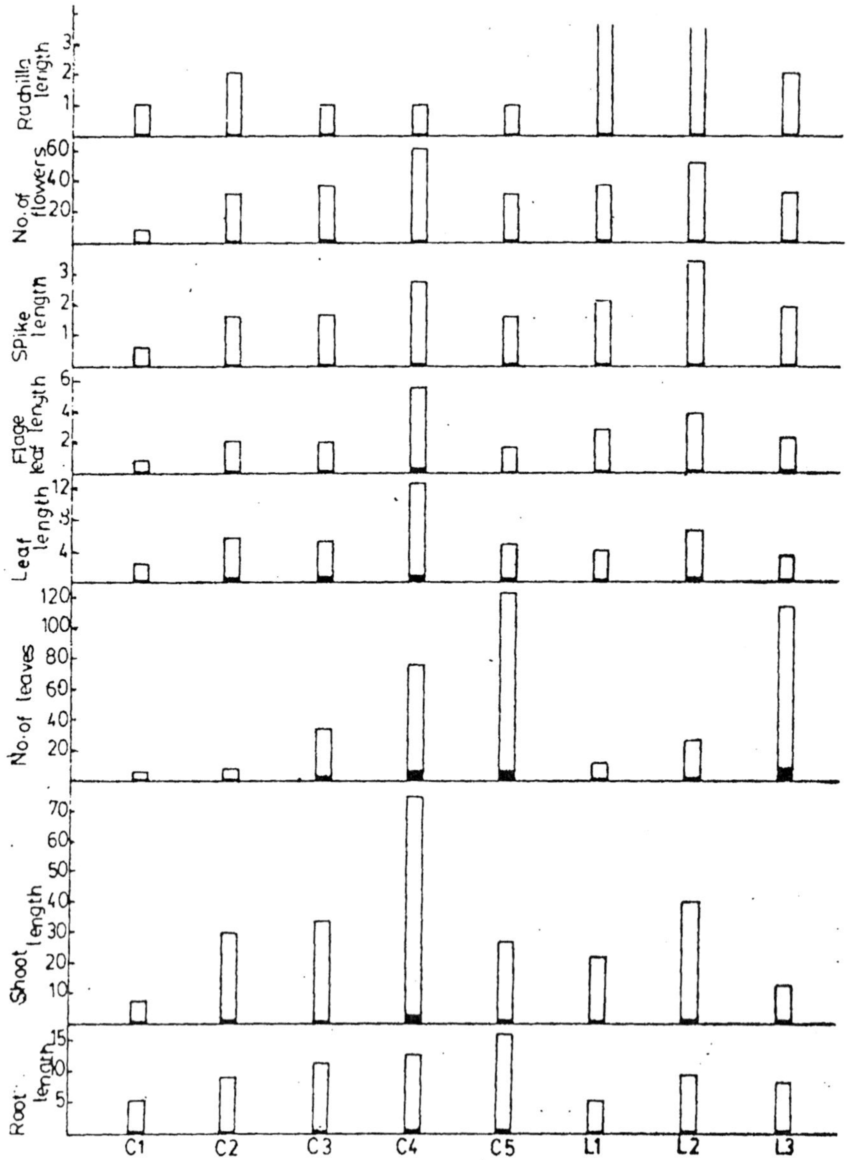

Fig. 8: Histograms showing variation in the mean (whole columns) and standard error (solid parts of the columns) among the forms for some of the morphological features.

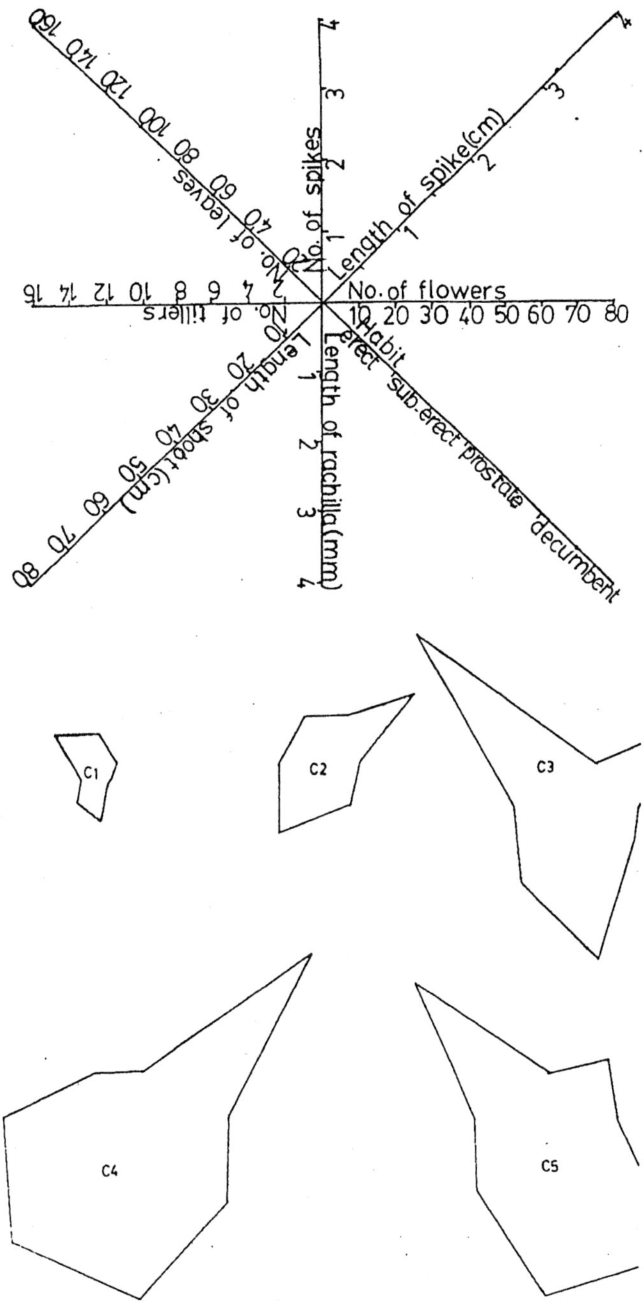

Fig. 9: Phytographs of the compact forms C1–C5 and laxforms L1-L3. Drawn as per the plan given.

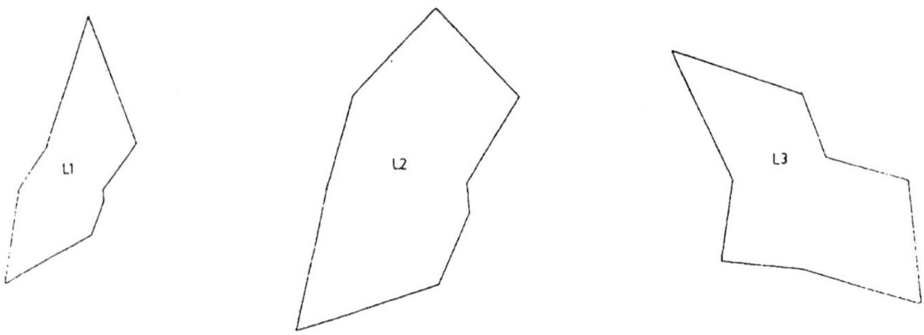

Fig. 10: Phytographs of the Lax forms L1–L3. Drawn as per the plan given

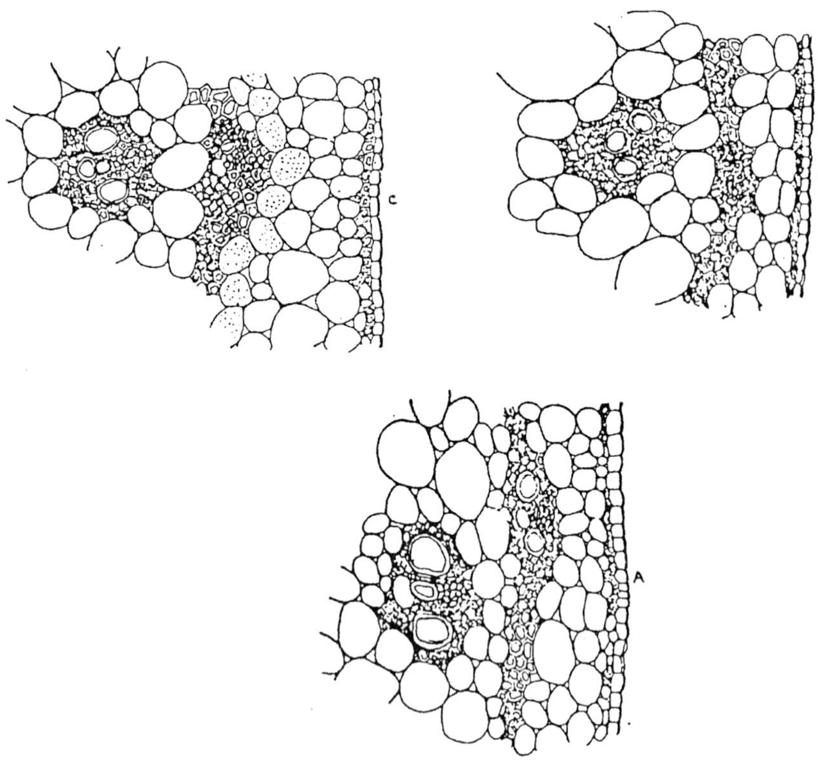

Fig. 11: Anatomy of the culm among the three forms: a—C4 discontinuous sclerenchymatous hypodermis and larger vascular bundles; b—22 continuous sclerenchymatous hypodermis; c—C5 discontinuous sclerenchymatous hypodermis and large portion of ground tissue having chlorenchymatous tissue and smaller vascular bundles.

observed in L3 (Plate 9). There was no evidence for structural alterations of the genome. Twenty PMCs were counted for chromosome pairing in these forms. The frequency of ring bivalents was found to be more than rods e.g., the mean chromosome pairing/cell at diakinesis was 21.5 rings, 1.5 rods and xta as 44.0 for C4 and 17.9 rings, 2.1 rods with xta of 39.0 for L2, 17.0 rings, 2.0 rods and xta of 36.0 for L3. Anaphase-I showed regular separation of the chromosome except 1–2 laggards in 40 chromosomes cytotype L2.

Pollen grains varied in size from 25.0 µm to 32 µm with an average of 30.80 µm in C4 and from 25 µm to 30.0 µm with an average of 27.00 in L2. The pollen-fertility was 90% in C4 while 70% in L2.

Seed Studies

Seed Dispersal and Seed Output

The seeds were light and wind-borne in all the forms. They were leached down by water during rains. A large number of seeds were seen around the ant holes (Plate 10a & b). Following Ridley (1930) and Vander Piji (1969), the seed dispersal in *D. aegyptium* can be described as the polychore type. The seed output varied among the forms with respect to the size of the spike and the number of flowers per spike. The highest values were noticed in C4 and the lowest in C1. The average seed output values for the different forms are as given in Table 4.

Seed Size, Weight, Colour and Shape

As it is evident from the Table 4 there are no marked differences in the size and weight of seeds among all these forms. Even regarding the shape, there was no difference between the two main groups. In general, the seeds were oblique, irregular in outline, slightly longer than

Table 4. Output, weight and diameter of seed in different forms of *D. aegyptium* complex

Form	Seed output	Seed weight (mg)	Seed diameter (mm)
C1	8.30	0.315	0.8259
C2	63.66	0.285	0.8407
C3	489.46	0.270	0.8222
C4	1967.29	0.320	0.8667
C5	753.09	0.295	0.8667
L1	50.89	0.365	1.0111
L2	319.85	0.410	0.9629
L3	659.71	0.440	1.0111

wide, somewhat compressed and rugose. C3 had lowest value for seed weight and size. The colour was orange brown in the compact forms and light grey in the lax forms (Plate 8a).

Seed Germination

Germination is epigeal (Plate 8b). Freshly collected seeds did not germinate readily because of dormancy. Hence a series of experiments were carried out to study the germination behaviour under diverse conditions such as dry storage, water submergence, water stress, salinity, and light (intensity and quality). Effect of growth regulators like IAA and Kn, scarification by sand and sulphuric acid, germination at different depths and different soils were also tried. There was no differential response in germination between compact and lax forms.

Seed Germination in Different Soils

Seed germination was high in the soil prepared by mixing sand, garden manure, clay and laterite in equal proportions. The order of germination percentage was sand < garden manure < clay < laterite < all mixed (Table 5).

Table 5. Seed germination (%) in different types of soils

Form	Type of soil				
	Sand	FYM (garden soil)	Black clay	Red literitic	All mixed
C1	0	0	4 ± 0	6 ± 1	9 ± 2
C2	0	2 ± 0	9 ± 2	8 ± 1	11 ± 1
C3	11 ± 0	4 ± 1	3 ± 1	6 ± 1	11 ± 2
C4	0	2 ± 0	8 ± 2	8 ± 2	14 ± 2
C5	4 ± 1	4 ± 0	6 ± 1	17 ± 3	23 ± 2
L1	2 ± 0	2 ± 1	4 ± 1	7 ± 2	20 ± 4
L2	0	6 ± 1	19 ± 4	16 ± 3	22 ± 3
L3	2 ± 0	7 ± 1	7 ± 1`	12 ± 2	19 ± 3

Reproductive Capacity

This plant reproduces by seeds and also by vegetative means. Reproductive capacity was calculated according to the formula given under Material and Methods. The reproductive capacity in the different forms was in the following order: C4 > C5 > L3 > C3 > L2 > C2 > L1 > C1 (Table 6).

Table 6. Reproductive capacities of different forms of *D. aegyptium* complex

Form	The average seed output	% germination	Reproductive capacity
C1	8.30	64	5.31
C2	63.66	82	52.20
C3	489.46	79	386.67
C4	1967.29	80	1573.83
C5	753.09	82	617.53
L1	50.89	66	33.59
L2	319.85	76	243.08
L3	659.71	78	514.57

Aggressive Capacity

Countings of surviving seedlings were made at successive intervals in 10 pre-marked permanent quadrats for C4 and L2. The average percentage survival of seedlings along with the values of aggressive capacities is given in Table 7. The percentage survival of seedlings was 9.63 and 9.03 for C4 and L2 respectively. Obviously, the values of aggressive capacity for C4 were relatively high, since this form had high seed output and high reproductive capacity.

Table 7. Aggressive capacities of C4 and L2 forms

Form	Average % survival of seedlings	Average reproductive capacity	Aggressive capacity
C4	9.63	1573.83	151.56
L2	9.03	243.08	21.95

Leaf Studies

Leaf Size and Stomatal Index

The leaf size in different forms follows the orders:

C4 > L2 > C2 > C3 > L1 > L3 > C5 > C1

The stomata are dumbell-shaped and arranged in row parallel to the midrib. Stomatal index showed marked differences in the upper epidermis, but remained uniform in the lower epidermis, of different forms (Table 8).

Chlorophyll Content of Leaves, Shoots and Inflorescences

The chlorophyll *a* and *b* content greatly varied between the compact and the lax forms, the former always showing higher quantities than

Table 8. **Leaf area (cm^2), length and breadth (cm), stomatal index, chlorophyll stability index (CSI) in all the forms of *D. aegyptium* complex**

Form	Leaf parameters			Stomatal index		Average CSI
	leaf area	leaf length	leaf breadth	lower epide-rmis	upper epide-rmis	
C1	0.35 ± 0.15	2.01 ± 0.11	0.21 ± 0.01	19.23	12.56	0.07
C2	1.85 ± 0.47	5.58 ± 0.34	0.38 ± 0.01	19.12	13.21	0.111
C3	1.73 ± 0.30	4.93 ± 0.21	0.44 ± 0.01	19.03	13.34	0.05
C4	3.76 ± 1.69	12.77 ± 0.67	0.67 ± 0.03	19.43	15.57	0.19
C5	0.38 ± 0.16	4.67 ± 0.18	0.47 ± 0.02	19.87	9.81	0.02
L1	0.93 ± 0.15	4.19 ± 0.14	0.36 ± 0.01	19.00	7.03	0.14
L2	3.54 ± 0.40	6.41 ± 0.25	0.48 ± 0.02	19.34	7.33	0.03
L3	0.39 ± 0.14	3.08 ± 0.12	0.42 ± 0.01	19.28	8.72	0.03

the latter. Generally, leaves showed higher chlorophyll content followed by stems and inflorescences. These results hold good irrespective of the age of the plants (Tables 9 & 10).

Chlorophyll Stability Index (CSI)

The Chlorophyll Stability Index values of all the forms were estimated and compared (Table 11). The data reveal that the prostrate forms, C5 and L3, had lower CSI values, thus showing higher drought-resistant capacity over others.

Phenology

Phenology was studied on the basis of field observations made on six major phases in the ecological cycle of the crowfoot grass, i.e. germination, vegetative phase, flowering phase, fruiting phase, dispersal of seeds and death of the plant. The data for all the forms are pictorially represented as phenograms (Figs. 12 & 13), following Sen (1981). Many seeds germinated with the onset of monsoon i.e. in the second week of June, and the active vegetative phase continued up to about the last week of August when the populations were luxuriant. The leaves attained their full size about one month after their initiation. The vegetative phase was followed by flowering and fruiting during August and November. However, both the prostrate forms had prolonged vegetative phase. They entered into flowering stage towards the end of September, and some of them could survive up to the month of February. The seeds obviously remained dormant in the soil until the next season, and germination resumed after the first monsoon showers. The species thus conforms to the life form, therophyte, of Raunkiaer (1934).

Table 9. Chlorophyll *a* of leaf, stem, and inflorescence of 30–60-, and 90-day old plants of all the forms of *D. aegyptium* complex

Form	Leaf			Stem			Inflorescence		
	30 days	60 days	90 days	30 days	60 days	90 days	30 days	60 days	90 days
C1	1.09 ± 0.02	1.39 ± 0.08	1.64 ± 0.09	0.74 ± 0.00	1.10 ± 0.07	1.27 ± 0.05	0.51 ± 0.02	0.68 ± 0.00	0.74 ± 0.01
C2	1.13 ± 0.03	1.48 ± 0.15	1.71 ± 0.15	0.78 ± 0.02	0.94 ± 0.02	1.39 ± 0.08	0.49 ± 0.04	0.65 ± 0.02	0.74 ± 0.00
C3	1.14 ± 0.02	1.46 ± 0.00	1.69 ± 0.00	0.76 ± 0.02	1.23 ± 0.15	1.47 ± 0.00	0.49 ± 0.00	0.73 ± 0.00	0.73 ± 0.01
C4	1.16 ± 0.01	1.62 ± 0.08	1.86 ± 0.08	0.86 ± 0.06	1.34 ± 0.20	1.80 ± 0.09	0.25 ± 0.00	0.49 ± 0.00	0.78 ± 0.02
C5	1.11 ± 0.07	0.95 ± 0.39	1.77 ± 0.09	0.69 ± 0.02	1.94 ± 0.15	1.80 ± 0.08	0.27 ± 0.01	0.49 ± 0.00	0.83 ± 0.02
L1	0.69 ± 0.04	0.98 ± 0.00	1.21 ± 0.00	0.51 ± 0.02	0.73 ± 0.00	0.97 ± 0.02	0.20 ± 0.01	0.25 ± 0.00	0.36 ± 0.01
L2	0.78 ± 0.04	0.96 ± 0.01	1.37 ± 0.09	0.51 ± 0.02	0.76 ± 0.01	0.95 ± 0.02	0.17 ± 0.00	0.22 ± 0.00	0.36 ± 0.01
L3	0.94 ± 0.10	0.99 ± 0.01	1.21 ± 0.00	0.57 ± 0.04	0.30 ± 0.01	0.99 ± 0.01	0.18 ± 0.01	0.28 ± 0.02	0.39 ± 0.01

Table 10. Chlorophyll *b* of leaf, stem, and inflorescence of 30–60-, and 90-day old plants of all the forms of *D. aegyptium* complex

Form	Leaf			Stem			Inflorescence		
	30 days	60 days	90 days	30 days	60 days	90 days	30 days	60 days	90 days
C1	0.45 ± 0.01	0.63 ± 0.02	0.68 ± 0.02	0.35 ± 0.01	0.55 ± 0.04	1.52 ± 0.76	0.20 ± 0.03	0.42 ± 0.02	0.43 ± 0.01
C2	1.01 ± 0.01	0.68 ± 0.02	1.00 ± 0.04	0.38 ± 0.03	0.46 ± 0.01	0.74 ± 0.01	0.24 ± 0.01	0.41 ± 0.03	0.41 ± 0.01
C3	0.78 ± 0.14	1.11 ± 0.00	1.43 ± 0.00	0.33 ± 0.02	0.50 ± 0.01	0.71 ± 0.01	0.26 ± 0.01	0.52 ± 0.02	0.44 ± 0.02
C4	0.82 ± 0.00	1.19 ± 0.12	1.51 ± 0.12	0.33 ± 0.00	0.74 ± 0.01	0.82 ± 0.02	0.12 ± 0.00	0.23 ± 0.01	0.25 ± 0.02
C5	0.49 ± 0.03	1.19 ± 0.15	1.53 ± 0.14	0.35 ± 0.02	1.42 ± 0.19	0.74 ± 0.07	0.13 ± 0.01	0.31 ± 0.03	0.23 ± 0.02
L1	0.36 ± 0.00	0.48 ± 0.00	0.09 ± 0.06	0.29 ± 0.03	0.47 ± 0.05	0.49 ± 0.01	0.14 ± 0.00	0.11 ± 0.01	0.15 ± 0.01
L2	0.34 ± 0.01	0.47 ± 0.01	1.07 ± 0.08	0.31 ± 0.01	0.41 ± 0.03	0.51 ± 0.02	0.14 ± 0.00	0.10 ± 0.01	0.14 ± 0.02
L3	0.23 ± 0.03	0.46 ± 0.03	0.95 ± 0.00	0.28 ± 0.02	0.60 ± 0.01	0.56 ± 0.04	0.11 ± 0.01	0.14 ± 0.01	0.13 ± 0.01

Table 11: Leaf area (cm²), length and breadth (cm), stomatal index, chlorophyll stability index (CSI) in all the forms of *D. aegyptium* complex.

Form	Leaf Parameters			Stomatal index		
	leaf area	leaf length	leaf breadth	lower epidermis	upper epidermis	average CSI
C1	0.35 ± 0.15	2.01 ± 0.11	0.21 ± 0.01	19.23	12.56	0.07
C2	1.85 ± 0.47	5.58 ± 0.34	0.38 ± 0.01	19.12	13.21	0.111
C3	1.73 ± 0.30	4.93 ± 0.21	0.44 ± 0.01	19.03	13.34	0.05
C4	3.76 ± 1.69	12.77 ± 0.67	0.67 ± 0.03	19.43	15.57	0.19
C5	0.38 ± 0.16	4.67 ± 0.18	0.47 ± 0.02	19.87	9.81	0.02
L1	0.93 ± 0.15	4.19 ± 0.14	0.36 ± 0.01	19.00	7.03	0.14
L2	3.54 ± 0.40	6.41 ± 0.25	0.48 ± 0.02	19.34	7.33	0.03
L3	0.39 ± 0.14	3.08 ± 0.12	0.42 ± 0.01	19.28	8.72	0.03

As the natural populations are generally subject to unexpected clearing, grazing, trampling and other disturbances, an uninterrupted study of the different phases of the life cycle of the two typical forms i.e. C4 and L2, was carried out under garden conditions. The results are summarized in Table 12. Four days were required for germination and six for the emergence of the first leaf in both the forms. There were slight differences in both the forms regarding other phenological events like third leaf emergence, tiller initiation, spike initiation, seed formation, dispersal of seeds, etc. The total life span for C4 and L2 was 157 and 164 days respectively.

The spikes were observed to be visited by ants and other insects, but their role in pollination is doubtful. Wind pollination was prevalent. Areas that were artificially cleared and irrigated were not taken into account while constructing the phenograms.

Table 12. Physical composition of the soils of all the forms of *D. aegyptium*

Factors % diameter of the particles in Parenthesis	Sites of occurence of C1–L3							
	I (C1)	II (C2)	III (C3)	IV (C4)	V (C5)	VI (L1)	VII (L2)	VIII (L3)
Gravel (more than 2 mm)	10	12	15	8	24	10	10	25
Coarse sand (2–0.2 mm)	28	29	25	22	32	26	20	32
Fine sand (0.2–0.02 mm)	32	30	36	38	12	31	34	13
Silt1 (0.02–0.002 mm)	8	17	16	18	10	20	22	8
Clay (less than 0.002 mm)	12	12	8	14	22	13	14	22

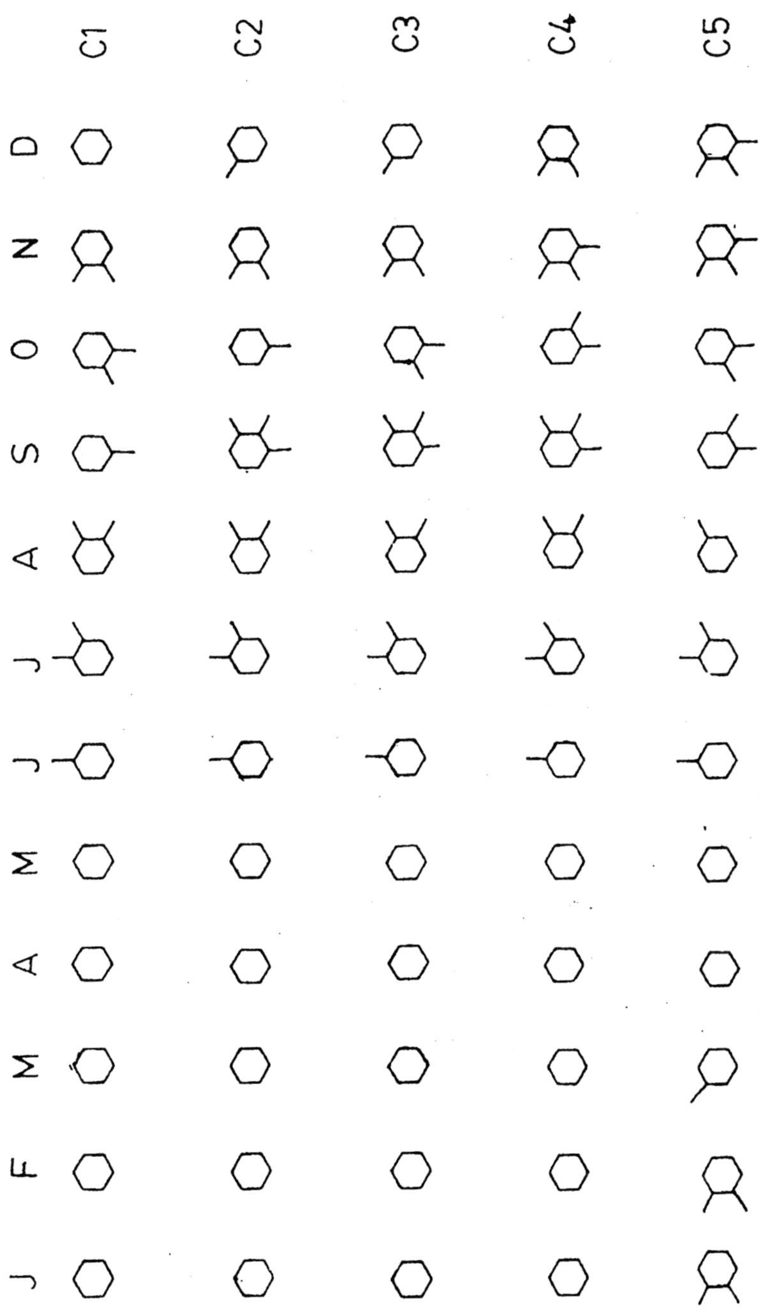

Fig. 12: Pictorial representation of phenograms of the compact and Lax forms.

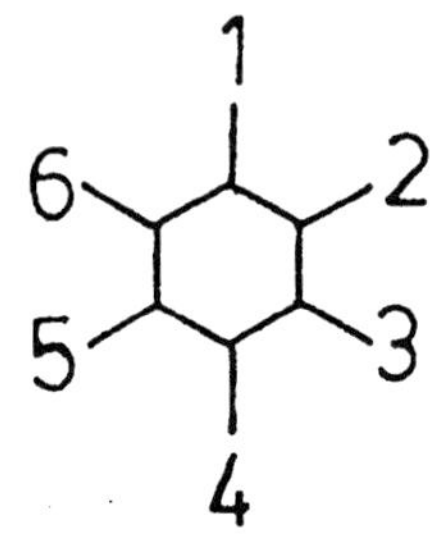

1 Germination
2 Vegetative
3 Flowering
4 Fruiting
5 Seed maturation
6 Death

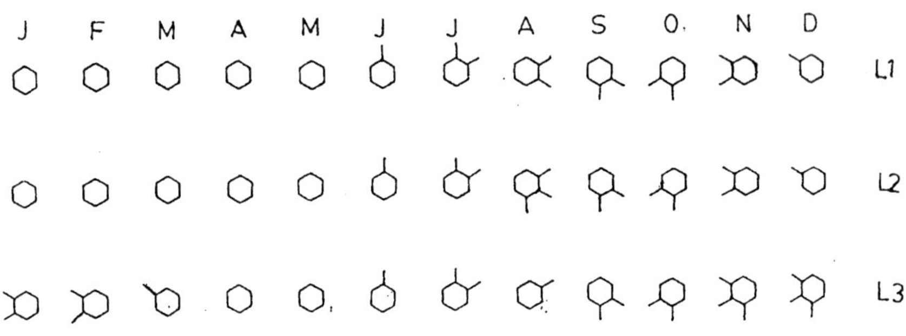

Fig. 13: Pictorial representation of phenograms of the compact forms.

Sociological Characters

The different sociological characters like density, basal area, IVI, etc., are calculated (Table 14). The association of species of *Chloris* and *Cymbopogon* is evident (Plate 10c).

Density

The total number of associated species as sampled from the quadrats of each form was 35. The relative density was more in C1 and L1. The relative density of the compact forms was less than that of the lax forms. In the relative dominance, based on basal area, C1 was superior to L1.

Table 13. Chemical composition of soils of occurrence of all the forms of *D. aegyptium* complex

Factors	Sites of occurrence							
	I (C1)	II (C2)	III (C3)	IV (C4)	V (C5)	VI (L1)	VII (L2)	VIII (L3)
Moisture content (%)	16.83	18.43	15.00	21.10	12.60	16.82	19.60	14.20
Water holding capacity	26.80	27.10	24.00	25.60	18.40	24.60	24.30	18.40
pH	7.60	7.50	7.80	7.80	7.80	7.50	8.10	7.90
Inorganic nitrogen (%)	0.03	0.03	0.20	0.25	0.25	0.21	0.20	0.22
Exchangeable potassium (%)	0.18	0.18	0.18	0.20	0.17	0.18	0.20	0.18
Available phosphorus (%)	0.001	0.001	0.001	0.001	0.001	0.001	0.001	0.001
Exchangeable calcium (m.e. %)	0.40	0.42	0.41	0.42	0.40	0.40	0.41	0.42
Exchangeable magnesium (m.e. %)	0.26	0.25	0.27	0.27	0.26	0.26	0.25	0.25
Carbonates (%)	0.02	0.02	0.02	0.02	0.02	0.02	0.03	0.02
Organic carbon (%)	0.42	0.40	0.41	0.44	0.41	0.42	0.44	0.40

Table 14. The different sociological features of all the forms of *D. aegyptium* complex.

Form	Relative Frequency	Relative Density	Relative Dominance	Importance Value Index
C1	35.7	89.4	36.9	162.0
C2	446.0	18.7	10.2	73.5
C3	39.7	21.9	8.7	70.3
C4	28.1	19.4	19.6	67.1
C5	32.8	14.5	43.1	90.4
L1	43.1	83.9	74.5	201.5
L2	30.9	34.7	28.0	93.6
L3	35.2	34.4	17.3	86.9

Basal Cover

The basal cover for all the different forms of *D. aegyptium* was always more than their associates, except where *Cymbopogon* dominated. Amongst these forms, C1 and L1 had higher values for basal cover.

IVI

The IVI values of grasses and other species were calculated. The polygraphs show the relative frequency, relative density, relative dominance and IVI of the forms of *D. aegyptium* and other species

(Figs. 14, 15). The IVI values were always high for all the forms of *D. aegyptium* except for C4. Between compact and lax forms, the latter excepting the prostrate forms, showed higher values.

Similarity Index

Similarity index as was worked out for the compact and lax groups was 0.63.

Soil and Plant Analysis

Soil Analysis

The soils collected from the respective sites of various forms of *D. aegyptium* were analysed for various physical and chemical properties. The data are summarized in the Tables 12 & 13. All the forms were found capable of growing in the alkaline soils. Significant differences were observed only in moisture content and pH in places where C4, C5 and L2 forms grew. The sites IV and VII as indicated in Table 12 registered slightly high organic carbon values where the luxuriant growth of C4 and L2 was observed. The other chemical properties showed minor differences among the different sites. The water holding capacity was more in the sites I and II where the C1 and C2 occurred in dense populations. As regards the physical properties, the sites V and VIII where the prostrate forms grew had more gravel and coarse sand, and less of fine sand, silt and clay, thus providing xeric conditions to the forms.

Estimation of Total Nitrogen, Crude Protein and Phosphates

Data on the biochemical components, total nitrogen, phosphates, and crude protein are given in Table 15. It is evident that C4 had higher crude protein in root, stem and flag leaf and so did L2 in leaf, inflorescence and seed. The total phosphate content was more in all the parts of C4 as compared to L2.

Biotic Factors

The different forms of *D. aegyptium* were grazed by animals. Inter-and intraspecific competition was not uncommon. This species was also observed to be infected by the fungus, *Ustilago* sp.

A graphical representation of IVI values of *D. aegyptium* (dotted line), and its associates (solid line). Plygraphic method showing sociological characters is given below.

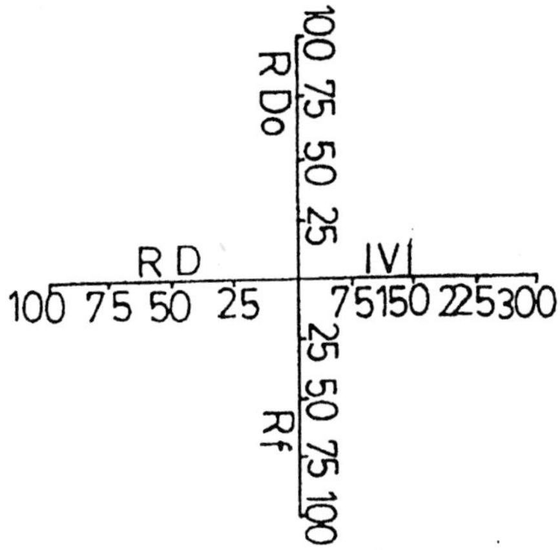

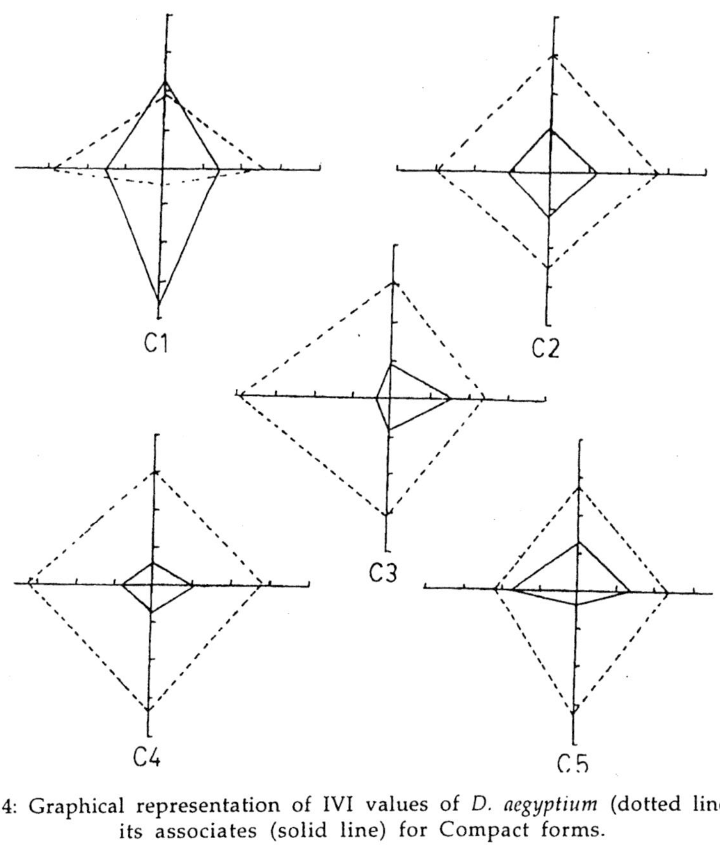

Fig. 14: Graphical representation of IVI values of *D. aegyptium* (dotted line) and its associates (solid line) for Compact forms.

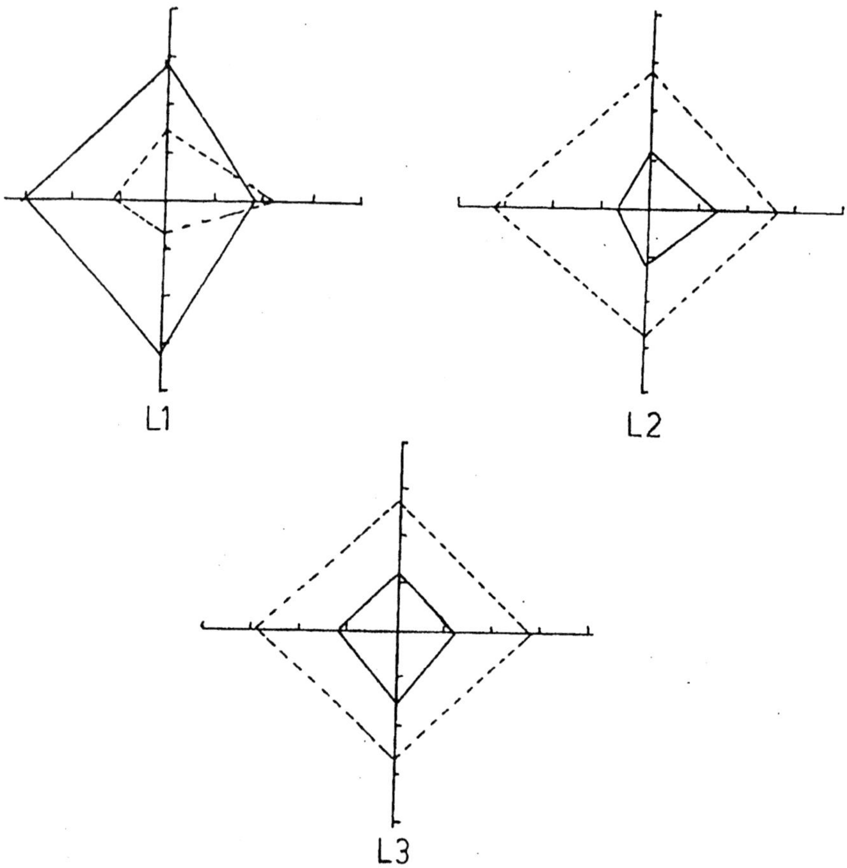

Fig. 15: Graphical representation of IVI values of *D. aegyptium* (dotted line) and its associates (solid line) for Lax forms.

Table 15. Total nitrogen, crude protein, and phosphate (%) of different plant parts in C4 and L2 of *D. aegyptium*

Factor	Form	Components					
		Root	Stem	Leaf	Flag leaf	Inflore- scence	Seed
Total Nitrogen	C4	1.16	0.57	0.08	1.63	0.56	1.96
	L2	0.57	0.52	1.39	1.04	1.19	2.34
Crude Protein	C4	7.26	3.59	0.53	10.19	3.50	12.25
	L2	3.54	3.24	8.70	6.48	7.43	14.61
Phosphate	C4	9.30	7.70	9.30	9.30	10.10	17.40
	L2	3.90	5.50	6.60	6.20	7.00	13.20

Discussion

It is understood from the culture experiments that the forms C1, C2 and C3 had more phenoplasticity than those of lax type in respect of length of culm, leaf, spike, number of tillers and spikes per plant, etc. However, these forms did not show any variation in rachilla length. Similarly, the leaves were sparsely hairy and the seeds orange brown in these forms. All of them had identical chromosome number (2n = 46). The length of the spike, a key character for identifying *D. aegyptium* (2–5 cm) varied in different manner from 1.0 to 4.5 cm. The culm height increases from C1 to C3, thus enabling them to exhibit a linear hierarchial development in sympatric populations. Such type of hierarchial development of plants in aged pastures was shown by Stern (1965) for *Trifolium subterraneum* and by Burdon & Harper (1980) for *Trifolium repens* from distinctly different environments. Gottlieb (1977) demonstrated genetic similarity of individuals at opposite extremes of the size distribution in the annual, *Stephanomeria exigua* and argued that the differences in plant size in an annual mainly reflect environmental factors. It was clear from these observations that the extent of morphological variation in compact group was more of quantitative than qualitative nature, as these three forms reverted to C4 type under uniform growth conditions of garden soil, they can be regarded as ecads.

The two compact forms, C4 and C5 and the lax form L2, however, maintained their identity in cultural experiments (Plate 11a to c). They differ in their growth habit, C4 being tall and erect and C5 exhibiting prostrate rosette-type of growth habit, typifies *D. aegyptium* whereas C5 represents a biotic ecotype of *D. aegyptium*. Though all the populations of *D. aegyptium* complex were observed on the campus grass land in the alkaline soils (pH 7.6–8.1), they had their own ecological preferences. For example, C3 and C4 were prevalent under high moisture conditions and C5 in drier habitats. Spatially their ecological niches were not far apart but only one to a few meters away from one another. The regular prevalence of prostrate population (C5) in the disturbed and dry habitats and the erect form (C4) in the protected and moist areas underscores the importance of soil moisture and biotic factors in the distribution of these forms.

There are several parallel instances documented for other grasses. For example, Kemp (1937) found that plant species from pastures with long history of grazing were markedly procumbent (see review by Watkin & Clements, 1978). Warwick & Briggs (1978) observed variation in growth habit in *Poa annua* from lawns and flower beds. In the lawns, prostrate plants were more due to regular mowing practices whereas erect plants were found at a selective advantage in flowerbed habitats. McNeilly (1981) also found similar, consistent differences in *P. annua* populations from closed and open habitats. The same was true of *Plantago major* from clipped lawns and unclipped roadsides (Warwick & Briggs, 1980). The established prostrate growth habit of C5 is traceable to heavy grazing pressures under drier habitats. Similarly the works of Hanson & Churchill (1961), Ramakrishnan (1960b), Jameson (1963) and Jain (1974) stress the above point further. It may be mentioned that the differentiation of morphological variability in this crowfoot grass in terms of ecads/ecotypes has not been dealt with by the earlier workers. The differentiation of these polymorphic forms is due to the evolution of phenotypic plasticity which enables the individuals to react during their development in an adaptive fashion to the specific conditions, and the direction followed by an evolving population depends on the genetic archetype of the species.

Within the lax group, as already mentioned, the forms L1 and L2 were non-stoloniferous, light green in colour and with profusely hairy leaves and light grey seeds. L1 mostly showed single spike whereas L2 showed a range of 1–3. On the other hand, L3 was prostrate, stoloniferous with 2–4 medium sized spikes; length of rachilla and hairy nature of leaf lamina and seed colour in all the lax forms was alike. According to Bor's key *D. aristatum* is non-stoloniferous with 0.5 to 2.0 cm long, solitary spike with rachilla produced upto 4 mm. A critical analysis of the morphological features of L1 and L2 clearly shows that they fit well with *D. aristatum* in essential features such as non-stoloniferous nature and rachilla length. However, they apparently differ from the existing description of *D. aristatum* by having remarkably longer spikes. It must be clarified here that the description of *D. aristatum* is incomplete so far as the range of spike length is concerned. Till now its known range is only 0.5 to 2.0 cm (see Bor, 1960). However, it is quite clear form the present study that the spike length in certain populations of L1 and L2 was between 1 and 4.5 cm. As this range of variation in spike length has not yet been documented for *D. aristatum*, several workers experienced intense difficulty in separating this taxon from *D. aegyptium* (see Sharma, 1984 and others).

Furthermore, that *D. aristatum*, which generally co-occurs with *D. aegyptium*, has often been confused with *D. aegyptium*, is also evident from the diverse chromosome number reports published under *D. aegyptium* (see Federov, 1974). The present study also brings into focus two highly useful characters which have not hitherto been mentioned in the literature. They are: hairiness of lamina and seed colour, which readily help distinguish the above two taxa. Consequently, Bor's key for the separation of these two species can be modified thus:

1. Plants stoloniferous, erect, prostrate to decumbent, tip of rachis shortly produced upto about 2.5 mm long, leaf lamina sparsely hairy with compact spikes; seeds orange brown .... ***D. aegyptium***
2. Plants not stoloniferous; erect to prostrate, tip of rachis upto 4 mm long (always more than 2.5 mm); leaf lamina profusely hairy; seeds light grey .... ***D. aristatum***

So far, *D. aristatum* has been known to be distributed in North-west India. Raju *et al.*, (1986) came across a few specimens of this species in Kurnool district (Andhra Pradesh, South-east India). According to them, it is extremely rare and restricted to Kurnool district only. However, the results of this investigation prove beyond doubt the common occurrence of *D. aristatum* in the coastal Guntur district of Andhra Pradesh, along with *D. aegyptium*. Its inconspicuous growth habit and short life cycle seem to be the possible reasons for *D. aristatum* having been not reported as a common grass species from southern part of India. In this connection, it is appropriate to quote Bor (1960), according to whom "there is one specimen of this plant at Kew, which according to Drummond is found in hard loam desert soil in North-west India and is usually passed over as *D. sindicum* or dwarf state of *D. aegyptium*."

In this study, it was found that both L1 and L2 were tetraploid races with the chromosome number being 2n = 40. Accordingly, they showed increased length of culm, leaf, number of tillers and spikes, etc., as compared with the description given for *D. aristatum* by Bor (1960). It seems likely that Bor's (1960) observations of *D. aristatum* might have been based on diploid cytotypes only. In general, polyploidy increases the growth vigour of the plants. As a result, such plants in grasses yield increased fodder to cattle. Sharma & Salam (1984) reported the existence of both diploid and tetraploid races in *D. aristatum* from Punjab plants. So, further meiotic studies in a few more populations is likely to help isolate the diploid cytotype in *D. aristatum*, inhabiting the present study site also.

What, then, is the status of L3? Morphologically, L3 appears to be intermediate between *D. aegyptium* and *D. aristatum*. It closely resembles *D. aegyptium* in the prostrate, stoloniferous branched growth habit and in having 2–4 spikes. Its affinity to *D. aristatum* is evident from hairiness of the leaf lamina, long rachilla and light grey seeds. Thus L3 represents a connecting link between *D. aegyptium* and *D. aristatum*, lending support to the view point that these species are not strictly reproductively isolated. The frequent occurrence of L3 in the study site, obviously indicates continuous interspecific hybridization of the two species, which has significantly contributed to the observed diversity of *D. aegyptium* complex.

Let us now understand the morphological complexity of the crowfoot grass from the view point of its cytological findings. From the diverse chromosome number reports published by several workers (see previous work), it is quite obvious that polyploidy must have been playing a crucial role in the morphological diversification of this taxon. From the existing reports, it is also evident that n = 9, 10 and 12 could be the basic numbers for this species, though n = 10 was given as the common basic number for the two tribes, Chlorideae (see de Wet, 1960; Pohl & Davidse, 1971; Tateoka, 1965a) and Eragrostoideae (see Darlington & Wylie, 1955; Carnaham & Hill, 1961). The chromosome numbers, n = 19, 20 and 23 as observed in this study, implicate the role of euploidy and aneuploidy in this species. In this context, it must also be mentioned that Davis (1966) had already provided evidence of apomixis in *D. aegyptium*. Accordingly, the various aneuploid series documented for this species must have been maintained by apomixis. In *Cenchrus ciliaris* as many as ten biotypes (see Donald, 1963; Darlington & Wylie, 1955; Jogindernath & Swaminathan, 1957; Patil *et al.*, 1961; Chandra Sekharan, 1961; Jagannath, 1963) have been identified. These irregularities in chromosome number among grasses are maintained by aposporous and facultative apomixis, e.g., *Pennisetum* spp. and *Cenchrus* spp. (see Fisher *et al.*, 1954; Synder *et al.*, 1955).

The meiotic finding in the species-complex in question showed the presence of bivalents at both diakinesis and metaphase-I. Such bivalent formation and diploid-like behaviour is genetically controlled (Jauhaur, 1975). The regular formation of bivalents at meiosis in various cytotypes of *D. aegyptium* as reported by Sachdeva & Ritakals (1981), Sharma & Salam (1984), and Bir *et al.* (1986), emphasizes genetically enforced preferential pairing resulting in meiotic integrity of the constituent genomes, leading to seed fertility. The gene controlled pairing behaviour is thus well established in this taxon. This explain the high level fertility observed in *D. aegyptium* (90%) in spite of the prevalence of polyploidy.

A perusal of the previous work shows the existence of diverse cytotypes with or without morphologic distinctness in *D. aegyptium* complex (Sachdeva & Ritakals, 1981; Sharma & Salam, 1984; Bir *et al.*, 1986). The extent of morphological variation seems to depend on whether these cytotypes are intraspecific or interspecific. The above authors have not assigned them to specific taxonomic units except in one case where Sharma & Salam (1984) designated three cytotypes under *D. aegyptium* and two under *D. aristatum*.

The two cytotypes, n = 23 and n = 20, observed in this study did show morphologic distinctness and could be assigned to *D. aegyptium* and *D. aristatum* respectively. The third one with n = 19, though possessing characters intermediate between the two species, remained a distinct phenotype. There are several instances in literature where the differences in chromosome number were accompanied by morphological differentiation. For example, in *Cenchrus ciliaris*, the cytotypes with 2n = 36 and 44, and exhibited distinctive morphological characters (see Jagannath & Raman, 1974). On the contrary, there are also several examples exhibiting a little or no morphological variation, though they are recognised as different cytotypes as in the following grass species: *Cynodon dactylon* (see Tripathy *et al.*, 1977); *Cynodon arcuatus* (see Mary & Malick, 1970); *Digitaria bicornis* (see Mary and Malick, 1971). Carr (1976) found in the species complex of *Calycadenia pauciflora-ciliosa*, sympatric races differing in chromosome number but morphologically very similar. On the contrary, Carr (1977) subdivided two closely related species, viz., *Calycadenia multiglandulosa* and *C. hispida*, into morphologically distinct allopatric races despite similarity in chromosome number and structure. So, within a group of species complex, morphological divergence in one line proceeds more rapidly than chromosomal change, while in another parallel line, the reverse can occur. Apparently both lines of evolution must be operating in *D. aegyptium* complex.

It may not be out of place to deal briefly with the ecological behaviour of the crowfoot grass based on the present findings relating to its seed germination, reproductive and aggressive capacities, anatomy, phyto-sociology and energy allocations, etc.

Seed output among the populations under study varied from 8.30 to 1967.29 seeds/plant. Among all the forms, C4 exhibited highest values owing to its luxuriant growth. The present data on seed weight (0.32 mg) and size (0.8667 mm) corroborate Sen's (1981) observations. This species has no efficient seed dispersal mechanism. As a result, the seeds get accumulated around the plants, resulting in clumped distribution of the populations. The mode of possible seed dispersal, according to Sen (1982), is by way of myrmecochory.

Germination experiments reveals that at least four-month storage period was necessary for the fresh seeds to germinate. In general, the germination behaviour of the seed did not show appreciable variation in the populations representing *D. aegyptium* and *D. aristatum*. The seeds submerged in water recorded poor germination, indicating the susceptibility of the species to water logging conditions. That the species is photoblastic is evident from the better seed germination results observed in light than in dark condition. Singhal and Sen (1981) observed seed germination in *D. aegyptium* to be as high as 90 ± 1.4 per cent, with seeds subjected to 5 minutes of acid scarification. Sand scarification of seeds in the present study gave similar results. They also noted that the increased salinity did not favour seed germination. In winter annuals, summer seed dormancy appears to be an adaptation which allows the species to persist in hot dry summer habitats (Ratcliffe, 1961; Newman, 1963; Thompson, 1970; Baskin & Baskin, 1971a; Janssen, 1974). Winter annuals growing in open habitats are not drought-tolerant and hence summer dormancy of seeds is an adaptation to the xeric habitats (Baskin & Baskin, 1971b). In the crowfoot grass, the seeds mature by November and remain dormant till the onset of monsoon.

The values of reproductive capacity showed great variation among the different forms of *D. aegyptium* in accordance with number of tillers, with spikes, length of spike, etc. (Vijaya Koteswari and Mary, 1986). These findings are in conformity with the views of Salisbury (1942), who drew attention to the broad correlations between the reproductive capacity, the seed characters of a species and its ecological status. The high aggressive capacity is an asset to the species to spread over a wide area for its seedling establishment. The reduced reproductive and aggressive capacities of *D. aristatum* as compared to *D. aegyptium*, were found to be responsible for the poor spread of this species.

As regards culm anatomy, the noteworthy difference between *D. aegyptium* (exemplified by C4, C5) and *D. aristatum* (exemplified by L2) relates to the sclerenchymatous hypodermis, which is discontinuous in the former and continuous in the latter. Overall, the anatomical findings such as the presence of sclerenchymatous tissue, extensive formation of ground tissue and numerous vascular bundles in *D. aegyptium* are traceable to its adaptation to xeromorphic conditions (Mary and Vijaya Koteswari, 1984). So, the species is able to thrive as a good colonizer in drier habitats. Chakrabarthy & Dilip Gupta (1981) recorded similar growth adaptations in *Euphorbia hirta, Boerhaavia diffusa* and *Cleome viscosa* under disturbed xerophytic environment.

The leaf characteristics such as chlorophyll content, stomatal index, Chlorophyll Stability Index (CSI), and hairiness were considered in this study because they not only help understand the degree of population differentiation but are apt to throw light on the adaptive behaviour of the species in question. The populations of *D. aegyptium* were consistently dark green in colour whereas those of *D. aristatum* were light green. The same was reflected in the relative amounts of chlorophyll *a* and *b*, which were found to be higher in *D. aegyptium* than in *D. aristatum*. The plants of drier habitats generally record low values of stomatal index as a means to reduce water expenditure. Accordingly, the stomatal index of upper epidermis in *D. aristatum*, which inhabits sandy soils was distinctly lower than in the widely distributed *D. aegyptium*. This is in conformity with the earlier observation made by Sharma & Salam (1984). CSI can be used as a parameter to measure drought tolerance of plants (Murthy & Mujumdar, 1962). The present data clearly indicate that the populations of either taxon under drier conditions registered low CSI values unlike the populations under mesic conditions, thus showing their drought tolerance capacity.

The distribution of hairs on leaves showed consistent variation between the two taxa and hence it can be used as important taxonomic criterion for separating them, as has already been pointed out. Further, the studies on leaf epidermis of *D. aegyptium* complex by Sharma & Salam (1984) revealed the presence of micro, macro hairs and prickels. According to them macro hairs were present on both surfaces of leaf in *D. aegyptium* and on adaxial surface only in *D. aristatum*.

The phenology of *D. aegyptium* shows that it completes its life cycle within a short period of about six months, i.e. from June to November, covering rainy season and part of winter. *D. aristatum* differ from *D. aegyptium* in having a somewhat shorter life cycle because it appears in the month of June and completes its life cycle two months earlier, i.e. October. However, the phenophases of the two species are such that they do favour of interbreeding between them. In both species, vegetative phase is shorter than reproductive phase, and so they can be considered r-selected.

The phytosociological data collected during the present study clearly indicate that *D. aristatum*, which registered high IVI values, is a poor competitor as compared to *D. aegyptium*. That is why it enjoys a limited range of distribution.

It was noted in this study that prostrate plants were more vigorous, channeling more energy into vegetative growth than erect plants, and had greater k-value. Conversely, erect plants diverted more energy to floral production, and thus showed higher r-value. A similar situation was reported by Warwick & Briggs (1978) in *Poa annua*. The differences in the reproductive effort between prostrate and erect forms can be related to their soil condition, and the biotic pressure they face.

The biochemical analysis of the various plant parts also reveals variation in nitrate, phosphate, and crude protein between *D. aegyptium* and *D. aristatum*. In addition, seed protein estimations reflect the existence of genetic differences between the two taxa. The biochemical work of Sachdeva & Ritakals (1981) also lends further support to this viewpoint.

Conclusion

These results bring out the fact that a very high amount polymorphicity/genetic variability exists among these natural populations of crowfoot grass complex in a single habitat. These populations do not belong to *D. aegyptium* alone but actually represent *D. aegyptium* complex comprising *D. aegyptium* L. Beauv, *D. aristatum* Link and taxa with morphological characters intermediate between these two species. The observed variability was judged due to environment polyploidy, hybridization followed by Natural selection. The high level of genetic variation discovered in these natural populations indicate that they have plenty of scope for evolution to occur.

From this study, it becomes evident that *D. aristatum*, which was thought to be restricted to North-west India (Bor, 1960) is occurring sympatrically with *D. aegyptium* in the present study site. The present study also brings into focus two highly useful characters viz., hairiness of leaf lamina and seed colour which readily help to distinguish the two taxa. In the light of the present data, a modification to Bor's key for the separation of these two species was also suggested.

The occurrence of intermediate forms (L3) suggest that the two species *D. aegyptium* and *D. aristatum* are closely related and barriers for interspecific hybridization do not exist.

Monitoring of the level of genetic variation through molecular analysis in this species complex, may find applicable in resolving disputes of taxonomic identities and interrelationships of these populations. The application of molecular marker technology i.e., RFLP, RAPD, QTLS and SNPs etc., would be useful to unravel the genetic diversity existing in the crowfoot grass. Further research work based on these lines of study are underway.

Acknowledgements

The authors thank Prof. T. Pullaiah, S.K. University, Ananthapur for extending invitation to contribute this article. We also thank Mr. K. Omkara Moorthy for computer assistance.

REFERENCES

Avdulov, N.P., 1928. Systematic Karyology of the family Gramineae. Proc. All-Russ. Botanists Congress, 65–67.

Banerjee, A.K., 1972. Certain aspects of weeds of rice fields. D. Phil. Thesis, University of Calcutta, Calcutta.

Baskin, J.M. and Baskin, C.C., 1971a. The possible ecological significance of the light requirement for germination in *Cyperus inflexus*. Bull. Torrey Bot. Club, **98:** 25–33.

Baskin, J.M. and Baskin, C.C., 1971b. Germination of winter annuals in July and survival of the seedlings. Bull. Torrey Bot. Club. **98:** 272.

Bir, S.S. and Sahni, M., 1987. Chromosomal and morphological variations in grasses of Punjab. J. Cytol. Genet. **22:** 12–22.

Bir, S.S., Sahni, M., Kanta, S. and Gill, G.S., 1987. Cytological investigations on some grasses from Punjab plain, North India. J. Cytol. Genet., **22:** 23–47.

Bor, N.L., 1960. Grasses of Burma, Ceylon, India and Pakistan, Pergamon Press, London.

Braun Blanquet, J., 1932. Plant Sociology: The Study of Plant Communities. Trans., revi. and ed. by Fuller, G.D. and Conarad, H.S. McGraw-Hill Book Co., New York and London.

Burdon, J.J. and Harper, J.L., 1980. Relative growth rates of individual members of a plant population. J. Ecol. **68:** 953–957.

Barna Ham, H.L. and Helen, D.H., 1961. Cytology and Genetics of forage grasses. Bot. Rev., **27:** 1–162.

Carr, G.D., 1976. Chromosome evolution and aneuploid reduction in *Calycadenia pauciflora* (Asteraceae) Evolution, **29:** 681–699.

Carr, G.D., 1977. A Cytological conspectus of the genus *Calycadenia* (Asteraceae). An example of contrasting modes of evolution. Am. J. Bot., **64:** 694–703.

Chakrabarthy, T. and Dilip Gupta, 1981., Morpho-histologic studies on three herbaceous species of railway track. Proc. Indian Academy of Sciences. B, **90:** 305–312.

Chandola, R.P., 1959. Cytogenetics of millets. Cytologia, **24:** 115–137.

Chandra Sekharan, P., 1961. Cytological studies on some South Indian grasses. Proc. 2nd Acad. Agric. Sci. Agri. Coll. and Res. Instt. Coimbatore, 63–74.

Chopra, S.L. and Kanwar, J.S., 1976. Analytical Agricultural Chemistry. Kalyani Publishers. Ludhiana-New Delhi.

Choudari, G.N. and Varshney, U.K., 1986. Soil salinity affecting growth and reproductive behaviour of *Dactyloctenium aegyptium* (Linn.) Beauv. Recent advances in environmental biology. Professor D.N.Rao Commemoration volume (ed.) Ambasht, R.S., BHU, Varanasi.

Christopher, J. and Abraham, D., 1974. Studies on the Cytology of South Indian grasses. II subfamily Eragrostoideae. Cytologia, **39:** 561–571.

Darlington, C.D. and Wylie, A.P., 1955. Chromosome Atlas of flowering plants. Allen and Unwin Ltd., London.

Darwin, C., 1859. The origin of species by means of natural selection.

Davis, G.L., 1966. Systematic embryology of the angiosperms. John Wiley & Co., New York.

De Wet, J.M.J., 1960. Chromosome numbers and some morphological attributes of various South African grasses. Amer. J. Bot., **47:** 44–49.

Donald, G. De Lisle, 1963. Taxonomy and distribution of the genus *Cenchrus*, Iowa State. J. Sci., **37:** 259–351.

Federov, 1974. Chromosome numbers of flowering plants. Printed in the Federal Republic of Germany.

Fisher, B.S. and Schewickerdt, 1941. Am. Natl. Mus., **10:** 47.

Fisher, W.D., Bashaw, E.C. and Holts, E.C., 1954. Evidence for apomixis in *Pennisetum ciliare* and *Cenchrus setigerus* Agron. J., **46:** 401–404.

Fogg, D.N. and Wilkinson, N.T., 1958. The calorimetric determination of phosphorus, Analyst, **83:** 406–414.

Gambel, G., 1957. Flora of the President of Madras. Vol.III. BSI Calcutta, **3:** 1273.
Gottlieb, L.D., 1977. Electrophoretic evidence and plant systematics. Ann. Missouri Bot. Gard., St. Louis.
Gould, F.W., 1960. Chromosome numbers in South Western grasses. Am. J. Bot. **47:** 873–877.
Gould, F.W., 1966. Chromosome numbers of some Mexican grasses. Can. J. Bot. **44:** 1683–1696.
Grime, J.P., 1979. Plant strategies and vegetation processes. John Wiley and Sons, Chichester, New York.
Grob, A., 1896. Beiträge Zur Anatomic der epidermis. der graminum lätter Bible, Bot., (I) 1–107.
Gunzel, F., 1912–1913. Bot. Jb., 49, Beiblatt. **108:** 52.
Hanson, H.C. and Churchill, E.D., 1961. The plant community. Reinhold Publ. Corporation, New York.
Harborne, 1973. Phytochemical Methods. A guide toModern Techniques of Plant Analysis, Chapman and Hall, London, 205–207.
Holm, L.G., Plucknett, D.L., Pancho, J.V. and Herbergeri, J.P., 1977. The world's worst weeds–Distribution and biology. East West Centre, University Press, Hawaii, Honolulu, 244–248.
Hooker, J.D., 1897. (reprinted 1961). Flora of British India. L. Reeve & Co. Ltd., England, Vol. VII, 295–296.
Jackson, M.L., 1967. Soil chemical analysis. Prentice Hall, New Delhi.
Jagannath, D.R., 1963. Microsporogenesis in relation to fertility in three species of *Cenchrus*. M.Sc. (Ag.) Dissen. Madras University (unpubl.) Abstract in Madras Agric. J., **51:** 362.
Jagannath, D.R. and Raman, V.S., 1974. A study of cytological variation in chromosomal races of two species of *Cenchrus*. Cytologia, **39:** 49–56.
Jain, N.K., 1974. Studies in two ecotypes of *Vernonia patula* (Dryand) Merrill. I. Morphology and Physiology. J. Indian Bot. Soc., **53:** 88–96.
Jameson, D.A., 1963. Responses of individual plants to harvesting. Bot. Rev., **29:** 532–594.
Janssen, J.G.M., 1974. Stimulation of germination of winter annuals in relation to micro-climate and micro-distribution, aecologia. **41:** 197.
Jauhar, P.P., 1975. Genetic control of diploid like meiosis in hexaploid tall fescue. Nature, **254:** 595–597.
Jogindernath and Swaminathan, M.S., 1957. Chromosome numbers of some grasses. Indian J. Genet., **17:** 102.
Kemp, W.D., 1937. Natural Selection within plant species. Journal of Heredity, **28:** 329–333.
Kershaw, A.K., 1973. Quantitative and dynamic plant ecology. The English Language Book Society and Edward Arnold (Pub.) Ltd., London.
Krishnaswamy, N., 1931. See Chandola, R.P. 1959.
Kumar, Joshi, M.C. and Ramesh Babu, V., 1971. Some factors influencing the germination of seeds in two desert grasses. Trop. Ecol., **12:** 202–208.
Larsen, K., 1963. Studies in the flora of Thailand 14. Cytological studies in vascular plants of Thailand. Dansk. Bot. Ark., **20:** 211–275.
Mary, T.N. and Malik, C.P., 1970. A new chromosome race in *Cynodon arcuatus*. J.S. Presl. exc. B. Presl. The Indian Forester, **96:** 607–608.
Mary, T.N. and Malik, C.P., 1971. Studies in some *Digitaria* species. Genet. Iber., **23:** 167–181.
Mary, T.N. and Vijaya Koteswari, M., 1984. Some ecological and anatomical studies of the crowfoot grass, *Dactyloctenium aegyptium* (L.) Beauv. Geobios New Reports. **3:** 17–20.

Mary, T.N., Vijaya Koteswari, M., Surya Kumari, K. 1992. Biosystematic Survey of *Dactyloctenium aegyptium* complex from Andhra Pradesh. J. Cytol Genet. **27:** 63–68.

Mary, T.N., Satyakeerthi, M.R.P. and Hanumantha Rao, N., 1996. Biosystematic survey of *Dactyloctenium aegyptium* complex from Andhra Pradesh. II *Dactyloctenium aristatum*. Link. Geobios New Reports **15(2):** 151–152.

McMeilly, T., 1981. Ecolypic differentiation in *Poa annua*, inter population differences in response to competition and cutting. New Phytol. **88:** 539–547.

Mehra, P.N. and Sharma, M.L., 1975. Cytological studies in some central and eastern Himalayan grasses. Cytologia, **40(1):** 441–467.

Metcalfe, C.R., 1960. Antony of the monocotyledons, Vol.I. Gramineae. Oxford. Clarendon Press.

Misra, R., 1968. Ecology Workbook. Oxford & IBH Publ. Co., New Delhi.

Moffett, A.A. and Hurcombe, R., 1949. Chromosome numbers of South African grasses. Heredity, **3:** 369–373.

Murthy, K.S. and Mujumdar, S.K., 1962. Modification of the technique for determination of chlorophyll stability India in relation to studies of drought resistance in rice. Current Science. **31:** 470–471.

Nelvigi, V.G., 1967. Ecology of graying effects on reproductive capacity of six forbs in the grass vegetation of Varanasi, Ph.D. Thesis, Benaras Hindu University, Varanasi, India.

Newman, E.I., 1963. Factors controlling the germination date of winter annuals. J. Ecol., **51:** 625.

Olarode, O., 1974. Chromosome counts in some Nigerian grasses. Cytologia, **39:** 429–435.

Pandeya, S.C., Puri, G.S. and Singh, J.S., 1968. Research methods in Plant Ecology. Asia Publishing House.

Parodi, L.R., 1946. Gramineas Bonariensis clave para La determinacion de las generos. Yenumaracion de las especies–Buenos Aires, Acme Agency, 1–112.

Patil, B.D. (Miss.) Vohra, S.K. and Joshi, A.B., 1961. Chromosome numbers in some forage grasses. Curr. Sci. **30:** 393.

Piper, C.S., 1942. Soil and Plant Analysis. University of Adelaide, Adelaide, Australia.

Pohl, R.W. and Davidse, G., 1971. Chromosome numbers of Costa Rican grasses. Brittonia, **23:** 293–324.

Prat, H., 1932. L Epiderme des graminees Etude anatomique et systematique Ann. Sci. Nat. Bot., Set. 10, **14:** 117–324.

Prat, H., 1934. Application du principe de subordination aux caracteres epidermiques chey les Graminees. Trans. Roy. Soc. Can., Ser. 3, **28:** 19–23.

Prat, H., 1936. La Systematique des Graminees. Ann. Sci. Nat. Bot., Ser. 10, **18:** 165–258.

Raju, R.R.V., Prasanna, P.V. and Pullaiah, T., 1986. *Dactyloctenium aristatum*. Link—A little known grass from South India. J. Ecom. Tax. Bot., **8:** 246–248.

Ramakrishnan, P.S., 1960. Studies in the autecology of *Euphorbia hirta* Linn. J. Indian Bot. Soc., **39:** 455–473.

Rao, P.N., 1981a. Weed flora of Guntur In: Proc. of Youth for ecodevelopment, NSS Workshop, Guntur, A.P.

Rao, P.N., 1981b. On becoming aware about our weeds. In: Proc. of Youth for ecodevelopment, NSS Workshop, Waltair, A.P.

Ratcliff, D., 1961. Adaptation to habitat in a group of annual plants. J. Ecol., **49:** 187.

Sachdeva, S.K. and Ritakals, 1981. Cytologic, morphologic and chemo-taxonomic studies in *Dactyloctenium aegyptium* (L.) Beauv. Complex. Ind. Aca. of Sciences. Proceedings Plant Sciences, **90:** 217–225.

Salisbury, E.J., 1942. The reproductive capacity of plants. G. Bell & Sons Ltd., London.

Sant, H.4., 1963. Effect of graying on the general herbage of BHU ground. A study of seed character and reproductive capacity of our weeds. Proc. Nat. Acad. Sci., India. **33B**: 355–360.

Sant, H.4., 1964C. Grazing susceptibility of number of grasses and forbs from the grazing ground of Varanasi, India. J. Range Mangt., **17**: 337–339.

Sant, H.R., 1962. A study of the reproductive capacity of some grasses and forbs of the grounds of BHU. Ph.D. Thesis, Banaras Hindu University, India.

Sant, H.R., 1964a. Ecological studies in *Dichanthium annulatum* Stapf. with special reference to reproductive capacity in relation to grazing. Proc. Natt. Inst. Sci., India, **34B**: 354–372.

Sant, H.R., 1965. Vegetative reproductive capacity of *Cynodon dactylon* Pers. in relation to biotic disturbances. Sci. and Cult., **31**: 192–193.

Sant, H.R., 1964b. Seasonal variation in coverage of selected grasses and forbs in relation to grazing intensities in India. J. range Mangt., **17**: 74–76.

Schefler, W.C., 1969. Statistics for the biological sciences, Addison Wesley Pub. Co., London.

Sen, D.N., 1981. Ecological approaches to Indian Weeds. Geobios International, P.O. Box 14, Jodhpur, India.

Sen, D.N., 1982. Environment and Plant Life in Indian Desert. Geobios International, Jodhpur, India.

Sharma, C.B.S.R., Behera, B.N. and Dash, S.K., 1978. A Ctyological study of some grasses from Orissa an Eastern coastal belt of India. Indian Academy of Sciences Proceedings. B., Plant Sciences, **87**:

Sharma, M.L., 1984. On the morphological differences between *Dactyloctenium aegyptium* and *Dactyloctenium aristatum* (Gramineae), Curr. Sci., **53**: 326–327.

Sharma, M.L. and Abdus Salam, 1984. Biosystematic Survey of the *Dactyloctenium aegyptium* complex (Gramineae) in Punjab plains I. Cytotaxonomy. Res. Bull. Punjab Univ., **35**: 139–146.

Singh, J.S. and Yadava, P.S., 1974. Seasonal variation in composition plant biomass and net primary productivity of a tropical grassland at Kurukshetra, India. Ecol. Monogr., **44**: 351–356.

Singhala, B.K. and Sen, D.N., 1981. Seed germination in some grasses of Indian desert. Forage Res., **7**: 27–30.

Sorenson, T., 1948. A method of establishing groups of equal amplitude in plant society based on similarity of species content. K. Danske Vidensk. Selsk., **5**: 1–34.

Stern, W.R., 1965. The effect of density on the performance of individual plants in subterranean clover swards. Aust. J. Agric. Res. **16**: 541–555.

Suryanarayana, S. and Krishna Swamy, N., 1948. An easy method of obtaining epidermal peels of grass leaves. Current Science, **17**: 151.

Synder, L.A., Herandery, A.R. and Warmke, H.E., 1955. The mechanism of apomixis in *Pennisetum ciliare*. Bot. Gay., **116**: 209–222.

Tateoka, T., 1965a. Contribution to biosystematic investigations of East African grasses. Bull. Nat. Sci. Mus., **8**: 161–173.

Thompson, P.A., 1970. Changes in germination response of *Silene secundiflora* in relation to the climate of its habitat. Physiol. Plant, **23**: 739.

Tripathy, R.C., Sachdeva, S.K. and Malik, C.P., 1977. Cytogenetic studies and evolutionary patterns in some *Cynodon* species. Biol. Zentrald., **96**: 423–435.

Vijaya Koteswari, M. and Mary, T.N., 1984. The effect of some growth regulators and pesticides on pollen germination in the crowfoot grass (*Dactyloctenium aegyptium* (L.) Beauv.) Indian bot. Reptr. **3(1)**: 32–33.

Vijaya Koteswari, M., 1990. An appraisal of variability in the natural populations of *Dactyloctenium aegyptium* complex (poaceae) Ph.D. thesis. Nagarjuna University.

Warwick, S.I. and Briggs, D., 1978. The genecology of dawn weeds. II. Evidence for disruptive selection in *Poa annua* L. in a mosaic environment of bowling green lawns and flower buds. New Phytol., **81**: 725–737.

Warwick, S.I. and Briggs, D., 1980. The genecology of lawn weeds. IV. Adaphic significance of variation in *Bellis perennis* L. as revealed in a transplant experiment. New Phytol., **85**: 275–288.

Watkin, B.R. and Clements, R.J., 1978. The effects of graying animals on pastures. Plant relations in Pastures. (Ed: Wilson, J.R.) CSIRO, E. Melbourne, Australia. 273–289.

Willis, J.C., 1966. A Dictionary of the Flowering Plants and Ferns (Revised by Airyshaw, H.K.) Cambridge University Press.

CHAPTER 2

PHYTOGEOGRAPHICAL INVESTINGATIONS ON PTERIDOPHYTIC FLORA OF SATPURA HILLS: AN UPDATED ACCOUNT

*S.M. Vasudeva**

Satpura Hills comprising of Pachmarhi plateau and its adjacent areas in Hoshangabad district, Tamia plateau and Patal Kot Valley in Chhindwara district are located in the Madhya Pradesh state of central India. Earlier Bir and Vasudeva (1972) had provided with the phytogeographical observations on 56 members of Pteriodophytes based on the field collections made by the author of the present paper from Pachmarhi region only. Later on more areas surrounding Pachmarhi plateau as Nag Dwari, Chintaman, Chittar Shalla, Pachham Dwar, Rori Ghat and Nandia in Hoshangabad district and Tamia plateau and Patal Kot Valley in Chhindwara district were explored and intensive and extensive collections were made for a number of years by the author. Further more, the records made by Dixit (1989) Panigrahi & Dixit (1966) and Tiwari (1964) from the region have also been taken into account. Thus a total of 78 members of ferns and fern-allies are recorded from the Satpura Hills (For details reference is made to Vasudeva & Bir, 1994). So, to make phytogeographic observations more comprehensive an updated account is presented herewith.

Phytogeographically Satpura Hills have special significance in the fact that these lie between the Himalayas in the north and Palni and Nilgiri Hills in the south of the Indian Sub-continent. So far, a total of 78 members of Pteridophytes have been recorded from this region. Though the number of species is not very large, yet there is a great abundance of the individuals of the majority of the species, which have wide distribution all around Satpura Hills in deep ravines and

* Department of Botany, Punjabi University, Patiala-147 002 (Punjab)

various dense forests. Tree ferns are not only conspicuous by their presence but also due to their abundance.

The author has collected 9 ferns-allies and 49 ferns from Satpura Hills and its environs. Some species earlier recorded and included here (marked thus*) could not be found in spite of extensive and intensive field surveys carried out for 10 consecutive years because due to cutting down of certain forests, extension of agriculture in *terai* (at lower levels) region and road building activities, these species must have been extirpated from the region or their record may be based on wrongly indentified specimens. Vouchers of all the earlier works on the Pteridophytes of the region were not available to the author for checking up of the materials although great efforts were made to locate them. Only some of the specimens from Satpura Hills were available for study at DD and I have rejected some of the recordings for the region. One such example is of *Pteris vittata* Linn. (Labeled as *Pteris longifolia* Linn.). This is recorded from Satpura Hills by Gamble (1892) on the basis of Duthie's specimen, which is actually *Blechnum orientale* Linn. I have no basis to reject certain other species from Satpura Hills unless and until their specimens are examined.

Information about the distribution of Satputa hills Pteridophytes within India is given in Table 1. For comparison sake, three main regions, namely, western and eastern Himalayas and south India are taken because of their richness of flora. The plains of northwestern India are generally devoid (only about half a dozen of members are recorded) of Pteridophytic vegetation because of scanty rainfall and prevailing dry climate. Mountains of western India also do not harbor vegetation rich in Pteridophytes. A scrutiny of distribution of data (Table 1) for various members shows the following interesting features:

a) Species common with the western Himalayas (areas west of Nepal): 52.56%
b) Species common with the eastern Himalayas: 74.35%
c) Species common with the entire Himalayan region: 83.33%
d) Species from Satpura Hills that grow both in the eastern and western Himalayas: 55.12%
e) Species common with south India: 87.17%

Nineteen species from Satpura Hills namely, *Lycopodium cernuum* Linn., *Angiopteris evecta* (Forst.) Hoffim., *Cheilanthes tenuifolia* (Burm.) Sw., *Microlepia speluncae* (L.) Moore, *Nephrolepis exaltata* (L.) Schott., *Dicranopteris linearis* (Burm.) Underwood, *Cyathea gigantea* (Wall. ex Hook.) Holttum, *Cyathea nilgirensis* Holttum, *Cyathea spinulosa* Wall. ex Hook., *Polystichum amabile* (Bl.) J. Sm., *Bolbitis virens* (Wall. ex Hook. & Grev.) Schott., *Tectaria leuzeana* (Gaud.) Copel., *Trigonospora ciliata* (Beneth.) Holttum, *Christella cylindrothrix* (Rosent.) Holtt., *Pseudocyclosorus falcilobus* (Hook.) Ching, *Asplenium cheilosorum* Kunze, *Asplenium normale* Don, *Pleopeltis macrocarpa* (Bory. ex Willd.) Kaulf., and *Leptochilus*

Table 1. Comparative distribution of Pteridophytic Flora of Satpura Hills in different Regions of India

Sr. No.	Name of the Species (Satpura Hills)	Western Himalayas	Eastern Himalayas	South India
	A. Fern Allies:			
1.	*Psilotum nudum* (L.) Beauv.	+	–	+
2.	*Lycopodium cernuum* Linn.	–	+	+
3.	*Equisetum debile* Roxb.	+	+	+
4.	*Equisetum ramosissimum* Desf. var. *altissimum* A. Br.	–	–	–
5.	*Selaginella ciliaris* (Retz.) Spring	–	–	+
6.	*Selaginella exigua* Spring	–	–	–
7.	*Selaginella radicata* (Hook. & Grev.) Spring	–	–	+
8.	*Selaginella rupestris* (L.) Spring	–	–	+
9.	*Selaginella bryopteris* (Linn.) Baker*	–	–	+
10.	*Isoetes panchananii* Pant & Srivastava	–	–	–
	B. Ferns:			
11.	*Ophioglossum gramineum* Willd.*	–	–	+
12.	*Ophioglossum nudicaule* Linn.	–	–	+
13.	*Ophioglossum reticulatum* Linn.*	+	–	+
14.	*Botrychium daucifolium* Wall. ex Hook. & Grev.	–	–	+
15.	*Botrychium lanuginosum* Wall.*	+	+	+
16.	*Angiopteris evecta* (Forst.) Hoffm.	–	+	+
17.	*Osmunda regalis* Linn.	+	+	+
18.	*Lygodium flexuosum* (L.) Sw.	+	+	+
19.	*Adiantum philippense* Linn.	+	+	+
20.	*Adiantum incisum* Forsk.	+	+	+
21.	*Adiantum capillus-veneris* Linn.	+	+	+
22.	*Cheilanthes bicolor* (Roxb.) Fraser–Jenkins	+	+	+
23.	*Cheilanthes anceps* Blandford*	+	+	+
24.	*Cheilanthes tenuifolia* (Burm.) Sw.	–	+	+
25.	*Pteris cretica* Linn.	+	+	+
26.	*Pteris biaurita* Linn.*	+	+	+
27.	*Pteris quadriaurita* Retz.	+	+	+
28.	*Microlepia speluncae* (L.) Moore	–	+	+
29.	*Sphenomeris chinensis* (L.) Maxon	+	+	+
30.	*Araiostegia pulchra* (Don) Copel.	+	+	+
31.	*Leucostegia immersa* (Wall.) Presl	+	+	+
32.	*Nephrolepis cordifolia* (Linn.) Presl	+	+	+
33.	*Nephrolepis exaltata* (L.) Schott	–	+	+
34.	*Nephrolepis radicans* (Burm.) Kuhn*	+	+	–
35.	*Dicranopteris linearis* (Burm.) Underwood	–	+	+
36.	*Cyathea nilgirensis* Holttum*	–	+	+
37.	*Cyathea spinulosa* Wall. ex Hook.	–	+	+
38.	*Cyathea gigantea* (Wall. ex Hook.) Holttum	–	+	+
39.	*Dryopteris cochleata* (Don) C. Chr.	+	+	+
40.	*Dryopteris sparsa* (Don) O. Kuntze	+	+	+
41.	*Polystichum amabile* (Bl.) J. Sm.	–	+	+

contd. ...

... *contd.*

Sr. No.	Name of the Species (Satpura Hills)	Western Himalayas	Eastern Himalayas	South India
42.	*Polystichum auriculatum* (Linn.) Pr.*	+	+	+
43.	*Tectaria leuzeana* (Gaud.) Copel.*	–	+	–
44.	*Tectaria macrodonta* (Fee') C. Chr.	+	+	+
45.	*Egenolfia appendiculata* (Willd.) J. Sm.	–	–	+
46.	*Bolbitis virens* (Wall. ex. Hook. & Grev.) Schott.*	–	+	+
47.	*Athyrium anisopterum* Christ*	+	+	+
48.	*Athyrium falcatum* Bedd.	+	–	+
49.	*Athyrium japonicum* (Thbg.) Copel.	+	+	+
50.	*Athyrium hohenackerianum* (Kze.) Moore*	–	–	+
51.	*Athyrium schimperi* Moug. ex Fee'	+	+	–
52.	*Athyrium filix-foemina* (L.) Roth*	+	+	–
53.	*Athyrium pectinatum* (Wall.) Presl	+	+	+
54.	*Diplazium esculentum* (Retz.) Sw.	+	+	+
55.	*Dryoathyrium boryanum* (Willd.) Ching	+	+	+
56.	*Hypodematium crenatum* (Forsk.) Kuhn	+	+	+
57.	*Ampelopteris prolifera* (Retz.) Copel.	+	+	+
58.	*Trigonospora calcarata* (Bl.) Holtt.*	+	+	+
59.	*Trigonospora ciliata* (Benth.) Holttum	–	+	+
60.	*Trigonospora sericea* (Scott. ex Bedd.) Holttum	–	–	+
61.	*Pronephrium repandum* (Fee') Holttum*	+	+	–
62.	*Pronephrium nudatum* (Roxb.) Chandra	+	+	+
63.	*Christella cylindrothrix* (Rosenst.) Holttum	–	+	–
64.	*Christella dentata* (Forsk.) Holttum	+	+	+
65.	*Christella parasitica* (L.) Lev.	+	+	+
66.	*Christella subpubescens* (Bl.) Holttum*	+	+	+
67.	*Christella quadrangularis* (Fee') Holttum	+	+	+
68.	*Pseudocyclosorus falcilobus* (Hook.) Ching*	–	+	–
69.	*Asplenium cheilosorum* Kunze	–	+	+
70.	*Asplenium normale* Don	–	+	+
71.	*Asplenium unilaterale* Lam.	+	+	+
72.	*Asplenium inaequilaterale* Willd.	–	–	+
73.	*Blechnum orientale* Linn.	+	+	+
74.	*Microsorum membranaceum* (Don) Ching	+	+	+
75.	*Pleopeltis macrocarpa* (Bory. ex Willd.) Kaulf.	–	+	+
76.	*Leptochilus decurrens* Bl.	–	+	+
77.	*Marsilea quadrifolia* Linn.*	+	+	+
78.	*Marsilea minuta* Linn.*	+	+	+

decurrens Bl. are common with the eastern Himalayas but are not recorded from the western part. These constitute about 24.35% of the Pteridophytic flora of the area. *Selaginella ciliaris* (Retz.) Spring. *Selaginella radicata* (Hook. & Grev.) Spring, *Selaginella rupestris* (L.) Spring, *Selaginella bryopteris* (Linn.) Baker. *Ophioglossum gramineum* Willd., *Ophioglossum nudicaule* Linn., *Botrychium dancifolium* Wall. ex Hook. & Grev., *Egenolfia appendiculata* (Willd.) J. Sm., *Athyrium hohenackerianum* (Kze.) Moore, *Trigonospora sericea* (Scott. ex Bedd.) Holttum, and *Asplenium inaequilaterale* Willd. growing in Satpura Hills, are also found in south India, but are not known from the Himalayas. These constitute 14.10% of the Pteridophytic flora. It is also interesting to note that *Equisetum ramosissimum* Desf. var. *altissimum* A. Br., *Nephrolepis radicans* (Burm.) Kuhn, *Tectaria leuzeana* (Gaud.) Copel., *Athyrium filix-foemina* (L.) Roth, *Athyrium schimperi* Moug. ex Fee', *Pronephrium repandum* (Fee') Holttum, *Christella cylindrothrix* (Rosenst.) Holtt., and *Pseudocyclosorus facilobus* (Hook.) Ching met with both in Satpura Hills and in the Himalayas and constituting about 10.25% of the flora are not recorded from south India. There are only two species as *Selaginella exigua* Spring and *Isoetes panchananii* Pant & Srivastava constituting only 2.56% of the total Pteridophytic flora of the region, that are found in Satpura Hills alone and are not recorded either from the Himalayas or south Indian mountains.

From the above comparisons it is clear that the Pteridophytic flora of Satpura Hills is quite interesting from phytogeographic point of view. It consists of a mixture of temperate and tropical elements. The region is remarkable in forming a meeting ground for species characteristic to the Himalayas and the south Indian mountains. Therefore, the vascular seedless flora of the region resembles both the Himalayas and south Indian elements since more than three-fourths of the members met within the area are common with both the regions. Further, it may be pointed out that the Pteridophytic flora of Satpura Hills is more akin to the south Indian members (87.17%) than the Himalayan ones (83.33%) although the percentage difference is very insignificant. This similarity appears to be correlated with the long rainly season and the subtropical to tropical climatic conditions with higher temperature prevailing at the central and south Indian mountains as compared to the comparatively cooler climate in the Himalayas.

REFERENCES

Bir, S.S. and Vasudeva, S.M., 1972. Ecological and Phytogeographical observations on the Pteridophytic Flora of Pachmarhi Hills (Central India) *Jour. Indian Bot. Soc.* **51:** 297–304.

Dixit R.D., 1989. Ecology and Taxonomy of Pteridophytes of Madhya Pradesh *Indian Fern Jour* **6:** 140–159.

Gamble, J.S., 1892. The Ferns of Pachmarhi and those of Mehendragiri. *Indian Forester* **18:** 55–57.

Panigrahi, G. and Dixit, R.D., 1966. New Records of Ferns for Madhya Pradesh. *Proc. Nat. Acad Sci. India* **36:** 135–144.

Tiwari, S.D.N., 1964. Ferns of Madhya Pradesh. *Jour. Indian Bot. Soc.* **43:** 431–452.

Vasudeva, S.M. and Bir, S.S., 1994. *Pteridophytic Flora of Pachmarhi, Tamia and Patal Kot in Central India.* Indian Fern Society Publication, Western Printers, Patiala.

CHAPTER 3

CYTOPALYNOLOGICAL CHARACTERISTICS OF FERN-ALLIES OF SATPURA HILLS (CENTRAL INDIA)

*S.M. Vasudeva**

Satpura Hills comprising of Pachmarhi Plateau, Tamia Plateau, Patal Kot Valley and their surrounding areas located within a radius of 200 km in the Madhya Pradesh state of Central India are unique in having a variety of rare members of fern-allies. The detailed cytopalynological investigations carried out on these members have brought to light many interesting facts which are discussed here. Earlier an account of chromosome numbers and evolutionary status of ferns and fern-allies of Pachmarhi Hills was provided by Vasudeva & Bir (1983) but there is no information about cytopalynology of fern-allies of Satpura Hills as a whole.

The arrangement of families and genera is based on the system adopted by Löve *et al.* (1977) in "Cytotaxonomical Atlas of Pteridophyta". Voucher specimens are preserved in the herbarium of the Punjabi University, Patiala, (PUN.).

Family: **Lycopodiaceae**

1. *Palhinhaea cernua* (L.) Franco & Vasc. ("Club moss")

Materials studied from Badkachar Forest and Fanshaw Khud (Pachmarhi) revealed the presence of 104 bivalents at metaphase-I (4x-aneuploid). The chromosome size is very small and the chromosome number is high. From South India several aneuploid/ polyploid

* Department of Botany, Punjabi University, Patiala-147 002 (Punjab)

numbers as n = 104, 110, 136, 156, 208, etc. have been recorded. The presence of n = 104 in materials studied from Satpura Hills is in conformity with that of Ninan (1958) from Trivandrum in South India. High haploid chromosome numbers in different cytotypes indicate their high polyploid nature. Polyploidy and hybridization seem to have played a significant role in the evolution and diversification of the individuals of the species.

Spores almost round trilete (tetrahedral), triradiate scar obscure, non-perinous, hyaline, distal ornamentation rugulate to nearly absent, proximal surface faintly undulating, size 27–30 × 19–30 μm.

Family: **Selaginellaceae**

2. *Selaginella radicata* (Hook. & Grev.) Spring ("Spike moss")

Materials investigated from Fairy Pool and Guha Jal (Pachmarhi), Denwa ravine (Tamia) and Chhindi (Patal Kot) have revealed the presence of 36 bivalents at metaphase-I in several spore mother cells. Based on x = 9 the present species shows the highest grade of polyploidy (Octoploid) ever recorded in the genus. Earlier, Kuriachan (1963) had recorded the occurrence of tetraploid cytotype of the species with n = 18, 2n = 36 form Kodaikanal in South India.

Heterosporous, microspores oval, trilete (tetrahedral), non-perinous, ectine verrucose, bright yellow, size 30–57 × 24–49 μm, megaspores oval, hemispherical, trilete, tetrahedral, non-perinate, yellow, size 129–220 × 122–190 μm.

3. *Selaginella exigua* Spring ("Small spike moss")

Materials used for this study were collected from Lady's Point below Big Fall (Pachmarhi) and Denwa ravine (Tamia). Counting from several preparations clearly indicates the presence of 10 bivalents in several SMCs at metaphase-I. It is a diploid sexual species based on x = 10 and has been worked out for the first time. Amongst the fern-allies and ferns from Satpura Hills, it shows the lowest chromosome number (n = 10).

Heterosporous, microspores small, more or less circular, trilete, tetrahedral, non-perinate, pale yellow, size 27–68 × 27–65 μm; megaspores large, almost circular, terilete, brown, non-perinous, size 209–258 × 290–236 μm.

It is interesting to note that the two worked out species of *Selaginella* from Satpura Hills are based on x = 9 and x = 10, respectively. According to Kuriachan (1963) aneuploid change of chromosome number (from 10 to 9) takes place through hybridization followed by chromosome doubling and accumulation of intrachromosomal changes are operative for speciation within the genus.

Family: **Isoetaceae**

4. *Isoetes panchnanii* Pant & Srivastava ("Quillwort")

Several populations collected from Reechh Garh, Army Lake and muddy pool *en route* Chhota Mahadeo (Pachmarhi) unequivocally showed the presence of 55 bivalents and a fragment during metaphase-I in a large number of SMCs. Both SMCs and chromosomes are very small sized. All these plants are decaploid (10X) based on x = 11 and show the highest grade of polyploidy ever recorded in the genus. Earlier, Pant & Srivastava (1965) have worked out a tetraploid cytotype of the species with 2n = 44 + 1 frag, from Rewa in Madhya Pradesh.

Heterosporous, megaspores large, black when wet, white when dry, trilete (tetrahedral), perisporiate, size 190–304 × 186–285 μm; microspores small, monolete, bilateral non-perinate and very rare in occurrence.

Family: **Equisetaceae**

5. *Equisetum debile* Roxb. ex Vauch. ("Horse tail")

Different populations from Jambu Dwip, Down Fall and Beeja Khud (Pachmarhi) and forest near Mahadeo (Tamia) have unequivocally revealed the existence of 108 bivalents in SMCs at diakinesis and metaphase-I. The same numbers have been reported from North Indian and South Indian populations of this "horse tail" (Bir, 1973).

Spores simple, globose, alete, yellow, perinous, provided with four elongated, spirally twisted hygroscopic elaters which help in keeping spores grouped together, size 38–61 × 38–61 μm.

6. *Equisetum ramosissimum* Desf. var *altissimum* A.Br. ("Scouring rush"):

All the populations from different localities as Jambu Dwip, Down Fall and Beeja Khud (Pachmarhi) and Mahadeo Forest (Tamia) showed the same chromosome number, n = 108. At diakinesis 108 bivalents and at anaphase-II 108 + 108/ 108 +108 chromosomes are countable. This is in conformity with the earlier reports from the Himalayas as recorded by Bir (1973).

Spores simple, globose, yellow, alete, perinate, provided with four elongated, spirally twisted hygroscopic elaters for keeping the spores grouped together, size 38–57 × 38–57 μm.

From the above account it is clear that both the species of *Equisetum* show the same haploid number (n = 108). So far only the same chromosome number (n = 108, 2n = 216) is recorded in all the worked out species of the genus throughout the world (cf. Löve *et al.*, 1977). This is a well-established case of paleoploidy and the genus is considered to be based on x = 9 and all the members of the genus exhibiting n = 108 are ancient polyploids evolved in the remote past. In the extant species of the genus neoploidy has not been established at all and if we consider x = 108 as the basic number, then all the species are to taken as diploids which is highly improbable. The uniformity of chromosome number (n = 108) is all the 15 (cf. Löve *et al.*, 1977) worked out species of *Equisetum*, especially in a genus which is perhaps 'the oldest living vascular plant known to Science' is in strange contradiction to the behaviour of the majority of other members of Pteridophyta. Amongst the fern-allies and ferns of Satpura Hills the "horse-tails"/ "scouring rushes" (*Equisetum* species) exhibit the highest grade (24X) of polyploidy. The peculiar palynological features of this genus (Equisetum) are that its spores are alete and with elaters. Presence of elaters is a unique characteristic of the genus that is nowhere found in Pteridophytes.

Family: **Psilotasceae**

7. *Psilotum nudum* (L.) Beauv. ("Whisk fern")

It is one of the most rare and difficult materials to be collected in the field and then worked out cytologically. Populations from Chhota Mahadeo, Handi Khoh, Duchess Fall and Pancy Pool (Pachmarhi) and Denwa ravine (Tamia) have clearly shown 104 bivalents at metaphase-I indicating the species to be tetraploid on a basic number x = 52. This is in conformity with some of the previous reports from South India. Often 1–2 fragments were invariably observed in different populations. Interestingly multivalents have been found to be entirely absent in the studied populations from Satpura Hills, although the possibility of this species being an autopolyploid was indicated earlier by Manton (1950). A triploid form of the species with 2n = 156 + 3 which is reported from South India is not recorded from Satpura Hills in Central India. Interestingly, diploid form with n = 52 has not so far been recorded from India.

Spores monolete, bilateral, plano-convex to convace-convex to biconvex, oblong, ellipsoid, hyaline to pale yellow, ectine spinulose, non-perinous, size 38–65 × 23–34 μm.

It is interesting to note that this species is considered to be one of the most primitive species of Pteridophytes yet it has monolete bilateral type of spores, which is a common feature of advanced pteridophytic members.

Concluding Remarks

Cytopalynologically amongst fern-alllies of Satpura Hills it has been observed that:

1. Highest chromosome number (n = 108) and highest grade of polyploidy (24X) have been seen in *Equisetum debile* and *E. ramosissimum* var. *altisimum*. Also these are possessing alete spores with elaters which are unique feature amongst pteridophytes. 2. Lowest chromosome number (n = 10) and ploidy (2X) amongst fern-allies is seen in *Selaginella exigua* and the highest chromosome number (n = 36) and ploidy (8X) in the genus *Selaginella* have been observed in *S. radicata* these are also characterized by being heterosporous in nature. 3. Smallest spores (27–30 × 19–30 mm) are observed in *Palhinhaea cernua* where as largest (190–340 × 186–285 mm) in *Isoetes panchananii*. The former is also unique by having the aneuploid (n = 104) number whereas the latter has the smallest sized chromosomes amongst fern-allies as also the highest (n = 55 + 1f) number and ploidy (10X) in the genus. 4. *Psilotum nudum* is unique by the fact that though most primitive amongst pteridophytes is having monolete (bilateral) type of spores which otherwise is an invariable feature of the advanced members. Further the Satpura populations with n = 104 + 1f (4X) are noteworthy by the presence of largest sized chromosomes and the absence of multivalents thus not being antopolyploids in nature. 5. There is no correlation between the chromosome number/ ploidy level and the size of spore mother cells (SMCs)/ spores in fern-allies.

REFERENCES

Bir, S.S., 1973. Cytology of Indian Pteridophytes. *In: Glimpses in Plant Research* (Eds. P.K.K. Nair *et al.*), Delhi **1:** 28–119.

Kuriachan, P.I., 1963. Cytology of genus *Selaginella*. *Cytologia*, **28:** 376–380.

Löve, A., Löve, D. and Pichi Sermolli, R.E.G., 1977. *Cytotaxonomical* Atlas of the *Pteridophyta*, Cramer, Vaduz.

Manton, I., 1950. *Problems of Cytology and Evolution in the Pteridophyta*. Cambridge University Press, Cambridge.

Ninan, C.A., 1958. Studies on the cytology and phylogeny of Pteridophytes. II. Observations on the genus *Lycopodium*. *Proc. nat. Inst. Sci. India*, **24:** 54–66.

Pant, D.D. and Srivastava, G.K., 1965. Cytology and reproduction of some Indian species of Isoetes. *Cytologya*, **30:** 239–251.

Vasudeva, S.M. and Bir, S.S., 1983. *Chromosome Numbers and Evolutionary Status of Ferns and Fern-Allies of Pachmarhi Hills (Central India)*. International Bioscience Monographs, vol. 14 Today & Tomorrow's Printers and Publishers, New Delhi.

CHAPTER 4

GENETIC DIVERSITY IN THE GENUS *PENNISETUM*

*M.V. Subba Rao, P.V. Arjuna Rao and V. Manga**

Pennisetum is one among the larger genera of the grass family, Poaceae. It comprises about 140 species, which are widely distributed throughout the tropics and subtropics. This genus includes a common cultivated species *Pennisetum glaucum* (L.) R.Br. (Pearl millet or bajra) which has traditionally been a component of the dry land agriculture. It is considered to be the most drought tolerant among the domesticated cereals. Apart from pearl millet, which is cultivated mainly for its grain, some of the wild species of this genus are used for forage, ornamental, soil-binding and thatching purposes. In addition, most of the wild species of this genus possess agronomically desirable characters such as perennial nature, apomixis and resistance to diseases of agronomic significance—rust, smut and ergot.

Owing to the cross-pollinated nature of the species of *Pennisetum*, enormous genetic diversity exists at intra-specific level within the cultivated species *P. glaucum* and also at the interspecific level. This genetic diversity has been exploited for the constant improvement of the agronomic traits in the cultivated species. The present article attempts to present a comprehensive account of this multifaceted genetic diversity in this genus.

The genus *Pennisetum* belongs to the sub-family *Panicoideae* and tribe *Paniceae*. Two other genera of this tribe with which *Pennisetum* shows close resemblance are *Setaria* and *Cenchrus*. The resemblances between *Pennisetum* and *Cenchrus* are so close that the morphological

* Department of Botany, Andhra University, Visakhapatnam 530 003.

variation of these two genera overlap and show nearly continuous variation suggesting the possibility of a frequent gene flow between them. Probably the prominent feature that distinguishes these two genera is the nature of bristles or spines fused at the base in *Cenchrus* but free in *Pennisetum*.

Five sections have been identified in the genus *Pennisetum* (Stapf and Hubhard, 1934). These are (i) *Gymnothrix*, (ii) *Eu-Pennisetum*, (iii) *Penicillaria*, (iv) *Hetero-stachya* and (v) *Brerivalvula*. The section *Gymnothirx* was further sub-divided into two sub-sections *Pleuristigma* and *Acrostigma*. The two cultivated species, pearl millet (*Pennisetum glaucum*) and napier grass (*P. purpureum*), are placed in the section *Penicillaria*, a name reflecting the penicillate anther tips present in these two species. Brunken (1977) included these two species under what he identifies as section *Pennisetum* (*Pennisetum* sect. *Pennisetum*).

The probable center of origin for pearl millet is generally agreed to be Africa. Various regions within African Ethiopia (Vavilov, 1950), Nigeria (Murdock, 1959), Western Sudan and Senegal (Harlan, 1971) and Sahara (Clark, 1962) have been suggested to be the likely areas of origin. Based on the available evidence, Appa Rao and de Wet (1999) concluded that pearl millet originated in West Africa some 4000 years ago from where it reached India (its secondary center of diversity) some 2000 years ago through eastern and southern Africa. The pattern of evolution of pearl millet corresponds well with that expected on the basis of 'multiple centers' concept of domestication (Porteras, 1976., Purseglove, 1976).

Among the various species/and sub-species within the genus, *P. violaceum* (Lam.) L.Rich. [=*Pennisetum* sub sps. *violaceum* (*monodii*)] is generally considered to represent the wild progenitor of pearl millet. The single morphological character, which unambiguously distinguishes the wild and the cultivated forms, is the nature of involucres at maturity- deciduous in the wild and persistent in the cultivated. The remaining morphological variation is largely of continuous nature (Bilquez and Leconte, 1969; Belliard *et al.*, 1980; Bramel-cox *et al.*, 1984; Marchais and Tostain, 1985).

Based on the system of classification for a domesticated plant species proposed by Hanna and de Wet (1971), the variation in pearl millet falls into three gene pools- primary, secondary and tertiary. The primary gene pool (PGP) includes all those taxa that naturally cross with the cultivated species i.e., diverse forms (races, varieties etc.,) of cultivated *P. glaucum*, the wild progenitor *violaceum* (*mondii*) and the weedy form, *stenostachyum*. The secondary gene pool (SGP) includes all those species that can be crossed with the cultivated one and produce weak hybrids with various degrees of sterility. *Pennisetum purpureum*

Schum. and *P. schweinfurthii* fall under this category (Hanna, 1979, 90; Hanna and Dujardin, 1982, 86). The tertiary gene pool (TGP) includes species, which cannot exchange genes with the cultivated one or which do so rarely but form completely sterile hybrids and nearly all the remaining wild species can be listed under this category (Harlan, 1975; Hanna 1987).

I. Morphometric Diversity

The maximum extent of morphological diversity is, probably, exhibited with reference to plant growth habit and shape of caryopses. While the former is more cognizable at the interspecific level, the later assumes significance at the intraspecific level within *P. glaucum*.

In the growth pattern, *P. clandestinum* appears to be rather unique within this genus because of its prostrate habit and many short internodes. *P. purpureum* stands on the other extreme in these features being the tallest among the species of this genus and with longer internodes. Only a few like *P. glaucum, P. violaceum, P. schweinfurthii, P. ramosum* and *P. polystachyon* are annuals while others are perennials. In the foliar morphology, *P. hohenackeri* and *P. villosum* are different from other species in having long, narrow, stiff and needle like leaves. Two other species viz., *P. mezianum* and *P. clandestinum* possess very small leaves (Plate 1a & b). The inflorescence is a well developed laxy panicle with deciduous spikelets/involucres in all the wild species (including *P. violaceum*) except in *P. clandestinum* where it is reduced to 2-4 spikelets (Plates 2 & 3). The cultivated species *P. glaucum* has a more compact panicle with persistent spikelets, which remain attached to the rachis even after complete drying. The panicle shapes within the cultivated pearl millet display considerable diversity and have been grouped into nine different forms (Fig. 1) (Mangat, Maiti and Khairwal, 1999).

The caryopses (grain) of the wild and weedy species are included within the glumes which contrasts with the excluded, exposed nature of the grain in *P. glaucum*. In the latter the grain shapes appear so diverse that they have been recognized into eight types (Fig. 2) (Mangat, Maiti and Khairwal, 1999). This diversity in the shape and size of caryopses has been so prominent that it was used as a criterion (Brunken, 1977) to describe the general trends in the world collections of pearl millet germplasm and it followed the most consistent geographic pattern (Appa Rao and de Wet, 1999). Of the various shapes described, four basic types were found in the world collections (Table 1) and these have been used as racial characteristics. On this basis,

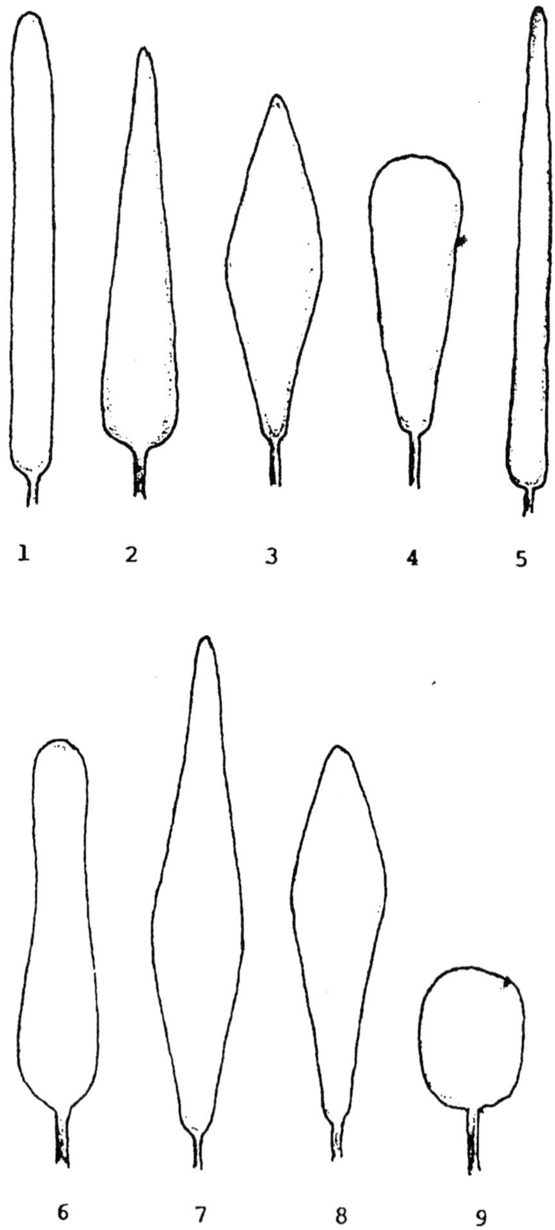

Fig. 1. Diversity in panicle shapes within the cultivated species (*P. glaucum*) (Modified after Mangat, Maiti & Khairwal, 1999): 1—Cylindrical; 2—Conical; 3—Spindle; 4—Club; 5—Candle Shape; 6—Dumb-bell; 7—Lanceolate; 8—Oblanceolate; 9—Globiose

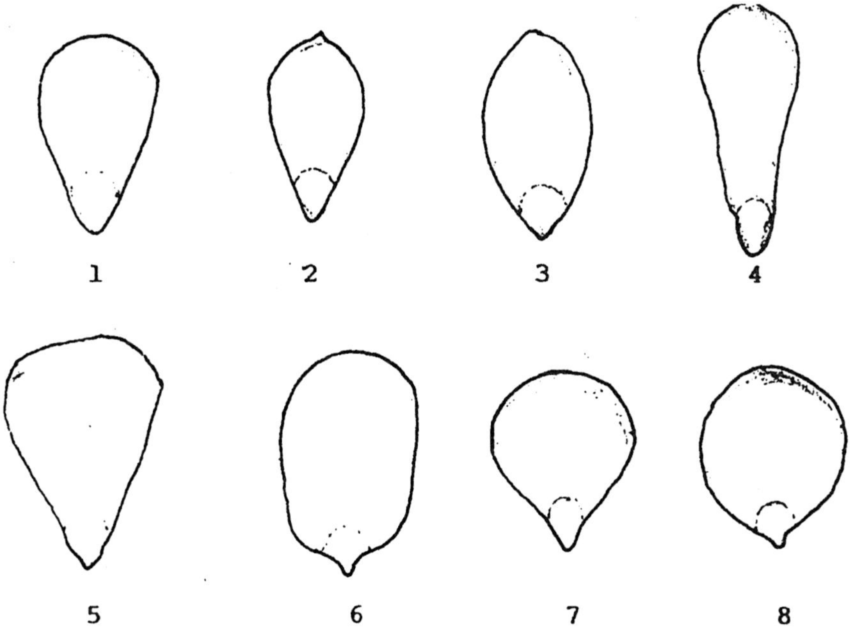

Fig. 2. Diversity in seed shapes within the cultivated species (*P. glaucum*) (Modified after Mangat, Maiti & Khairwal, 1999): 1—Obovate; 2—Lanceolate; 3—Elliptical; 4—Cuneform; 5—Pyramidal; 6—Hexagonal; 7—Pyriform; 8—Globular.

four races within *P. glaucum* have been classified (Appa Rao and de Wet, 1999). These patterns of racial variation were considered as suggestive of several independent domestications (Porters, 1976). These, together with geographic and ethnographic isolation as well as free out-crossings among different types consequent upon migration/ selection might be responsible for the present day diversity in the seed morphology.

The genetic diversity among related plant populations is generally assessed through multivariate analyses. Among these multivariate procedures, D^2 analysis has been very extensively used in pearl millet (Updhyay and Murty, 1971; Mukherji *et al.*, 1981; Andre Savery and Prasad, 1995). Of the different genetic stocks representing various ecological conditions, African and American stocks were relatively close to one another when compared to stocks from Asia and Australia. The Indian stocks showed distinct differences from all other geographic groups, especially for earliness and shorter height (Khairwal and Sultan Singh, 1999). Cluster analysis and Principal Component Analysis (PCA) have been applied to several pearl millet populations grown in Africa to understand the nature and degree of genetic

Table 1. Variation in seed shape and size used for racial differentiation with pearl millet (*P. glaucum*)

Race	Size			Shape	
	Length (mm)	Width (mm) dorsal-measure	Depth (mm)	Frontal and/profile view measure	Cross-section
Typhoides	2.5–5.5	1.5–3.0	1.2–2.4	Obovate	Obtuse and terete
Nigritareus	3.0– 5.0	1.7– 2.5	1.5–2.2	Elongate with usually truncate apex	Angular
Globosum	Length and width nearly equal		>2.4	Spherical	Terete and obtuse
Leonis	3.8–6.3	1.9–2.5	1.9–2.5	Acute/oblanceolate/terete Acute apex with remnants of stylar base	

divergence in terms of quantitative characters. Days to flower, plant height, stem diameter, primary panicle length, grain yield were the major traits contributing to diversity among the land race populations.

II. Anatomical Diversity

Very little information is available on the anatomical features in this genus. Foliar epidermal patterns in the cultivated and 11 wild species (*P. violaceum, P. purpureum, P. mezianum, P. hohenackeri, P. setacium, P. divisum, P. orientale, P. clandestinum, P. villosum, P. squamulatum* and *P. polystachion*) revealed no greater qualitative variation in the structure of the stomata, presence and distribution of long cells as well as short cells and the dermal appendages such as macrohairs, microhairs and prickle hairs. Only minor differences like absence of microhairs in *P. americanum, P. purpureum, P. divisum* and *P. clandestinum* have been observed. The cultivated species *P. americanum* showed very much restricted distribution of stomata compared to the wild species. Whatever little variation in the foliar epidermal patterns is known in this genus, it pertains mostly to quantitative aspects and/or in the distribution of the various characters on the two epidermal surfaces. The overall assessment that could be made based on the limited information is that the foliar epidermal characters are nearly uniform in at least 12 species of *Pennisetum* and exhibit no or negligible diversity (Sujatha, 1984).

III. Reproductive Diversity

A) Palynological Diversity

The importance of diversity in pollen morphology as an expression of the geographical distribution, genomic constitution, taxonomy, phylogeny and evolution of plants is being increasingly demonstrated in the recent years. Available information indicates greater diversity at the interspecific level than at the intraspecific level in the cultivated species. However, even this palynological variation is more of quantitative nature, being reflected in characters such as number of apertures, diameter of the pollen grain, diameter of the aperture, exine thickness etc. More striking among these characters is the number of apertures; nearly almost all the polyploid species studied showed diporate condition whereas the diploids (*P. glaucum, P. violaceum, P. schweinfurthii, P. ramosum* and *P. hohenackeri* were invariably monoporate. Some of the polyploids such as *P. setaceum* and *P. villosum* showed a variation from 0–4 in the number of apertures (Manga, Sujatha and Subba Rao, 1991).

Based on the pattern of exine ornamentation with special reference to the spinule distribution as observed under the scanning electron microscope, three groups can be recognized i) species with isolated spinules such as in *P. purpureum* and *P. setaceum*, ii) spinules arranged in groups as in *P. divisum, P. orientale, P. villosum, P. squamulatum* and *P. polystachion* and iii) mixed pattern with the above two types of arrangements as in *P. americanum, P. violaceum, P. mezianum, P. hohenackeri* and *P. clandestinum*.

B) Diversity for Male Sterility

Genetic diversity for male sterility has profound importance in breeding programs. Male sterility in pearl millet was first reported by Kajjari and Patil, (1956). The first male sterile line (Cytoplasmic) developed was Tift 23 (Burton, 1958) in USA which was a semi dwarf with good tillering capacity. This character was introduced into Indian variety 'Gahi' by repeated backcrosses and two male sterile lines Tift 18 A and Tift 23 A were developed (Burton, 1965a, b).

Two more cytoplasmic male sterile lines (L66A and L 67A) were developed at Punjab Agricultural University by Athwal (1961) which were different from Tift 23 in their cytoplasmic backgrounds; subsequently several male sterile lines were also developed (Table 2) by using /transferring these primary sources (A1 of Tift 23, A2 of L66 and A3 of L67). Appadurai *et al.*, (1982) developed genetic male sterility

Table 2. Male sterile types in pearl millet.

Line	Cytoplasm type	Origin
T23	A1	USA
L66	A2	Africa
L67	A3	Africa
L101	A1	Africa
L102	A1	Africa
L103	A2	India
L104	A1	USA
L105	A1	India
L106	A1	India
L107	A1	Africa
L108	A3	Africa
L109	A2	Africa
L110	A1, A2, A3	India
L111	A1	India
PT732	?	India
?	Different from A1, A2, A3	Africa

Source: Gill, 1991

A1. Genetic constitution: ms1 ms1 MS2 MS2 MS3 MS3.

A2. Genetic constitution: ms1 ms1 ms2 ms2 MS3 MS3.

A3. Genetic constitution: ms1 ms1 ms2 ms2 ms3 ms3.

system at Tamil Nadu Agricultural University, Coimbatore, Inida. The relationship between the cytoplasmic backgrounds of these Tamilnadu sources and those of A1, A2 and A3 remains to be investigated. Another source of cytoplasmic male sterility, which was shown to be different from A1, A2 and A3 background, was developed by Marchais and Tostain (1983) using Senegalese wild germplasm. Several of the wild species of *Pennisetum*, especially the polyploids, show varying degrees of sterility (both male and female) but no clear case conferring to the 'male sterility' type has been reported in any of them.

C) Apomixis and Polyembryony

Apomixis, in a broad sense, includes all asexual methods of reproduction but in a strict sense means agamospermy or formation of seed without fusion of gametes. Facultative agamospermy is very common among the polyploid species of *Pennisetum* and an extreme case of asexual reproduction by vegetative propagules, like bulbils has also been reported in *P. polystachion* (Nair and Pillai, 1969). While confirmed cases of apomixis have been reported in several polyploid cytotypes of wild species viz., *P. orientale* (Narayan, 1954, 62; Patil et al, 1962), *P. massaicum* (D'e ray and Reddy, 1968; Shantamma and Narayan, 1977a, b) *P. pedicellatum*, *P. setaceum* and *P. villosum* (Avdulov,

Plate 1: A & B Diversity in the Morphology of *Pennisetum* species:
A: 1—*P. americanum* (= *P. typhoides*); 2—*P. violaceum*; 3—*P. purpureum*; 4—*P. mezianum*; 5—*P. hohenackeri*; 6—*P. setaceum*
B: 1—*P. divisum*; 2—*P. orientale*; 3—*P. clandestinum*; 4—*P. villosum*; 5—*P. squamulatum*; 6—*P. polystachion*.

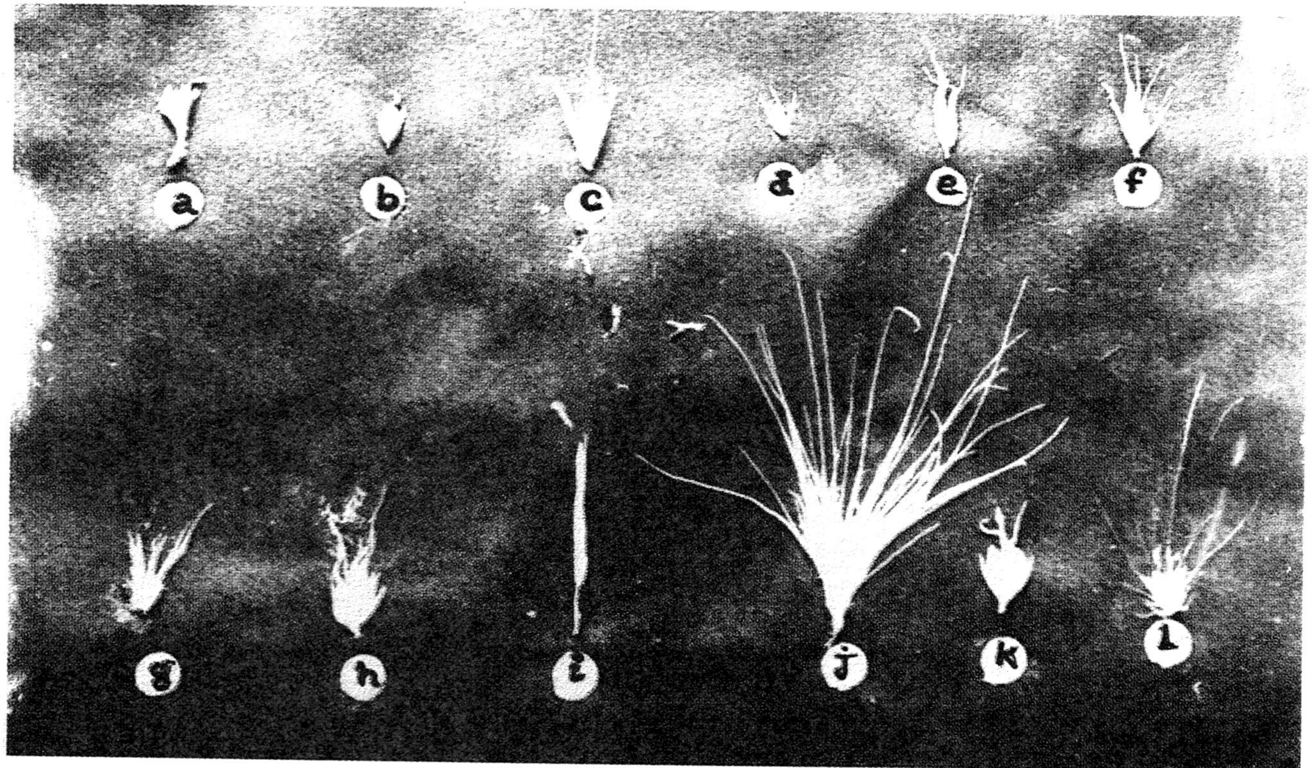

Plate 2: Diversity in the Panicles of 12 species of *Pennisetum*: a—*P. americanum* (= *P.typhoides*); b—*P. violaceum*; c—*P. purpureum*; d—*P. mezianum*; e—*P. hohenackeri*; f—*P. setaceum*; g—*P. divisum*; h—*P. orientale*; i—*P. clandestinum*; j—*P. villosum*; k—*P. squamulatum*; l—*P. polystachion*.

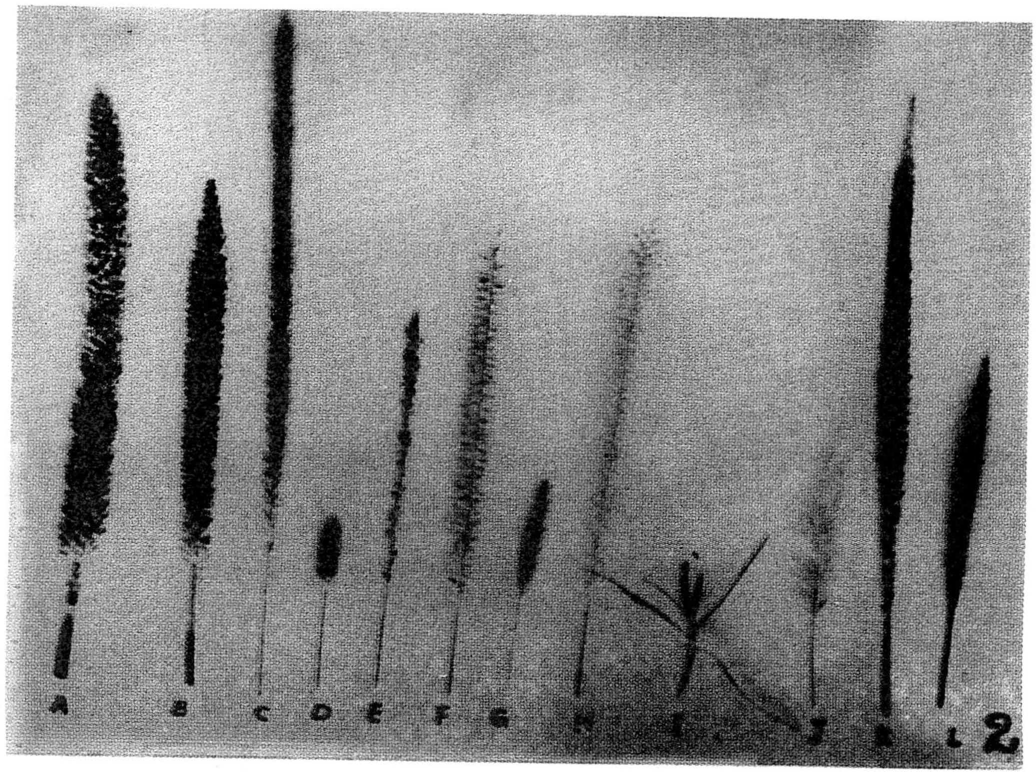

Plate 3: Diversity in the spikelet nature in 12 species of *Pennisetum*: a—*P. americanum* (= *P.typhoides*); b—*P. violaceum*; c—*P. purpureum*; d—*P. mezianum*; e—*P. hohenackeri*; f—*P. setaceum*; g—*P. divisum*; h—*P. orientale*; i—*P. clandestinum*; j—*P. villosum*; k—*P. squamulatum*; l—*P. polystachion*.

1931; Hrishi, 1952; Nunez, 1952; Narayan, 1962; Simpson and Bashaw, 1969; Chatterji and Pillai, 1970; Rangaswamy, 1972), apospory has been suggested in diploid *P. ramosum* (Narayan, 1954, 62), *P. clandestinum* (Narayan, 1951, 55) *P. flaccidum* (Chatterji and Timothy, 1969a) and in hexaploid *P. setaceum* (Simpson and Bashaw, 1969).

Multiple embryo sacs and polyembryonic caryopses have been reported in *P. clandestinum* (Narayan, 1958), *P. massaicum* (Shantamma and Narayan, 1977b), *P. orientale, P. setaceum* and *P. villosum*, (Narayan, 1951) as well as in *P. glaucum* (Powell and Burton, 1968; Powell *et al.*, 1975; Hanna *et al.*, 1976).

IV. Chromosomal Diversity

Chromosomal diversity is considered to reflect differences in the source of genetic variation while morphological, physiological and biochemical differences reflect differences in the products of gene action, modified by environmental influences (Stebbins, G.L., 1971). Though pearl millet possesses several desirable characteristics which make it a highly suitable organism for chromosomal analysis (Jauhar, 1981), the information on this aspect in the wild species is rather limited. The diversity at the chromosomal level is either numerical or structural or both.

The structural diversity includes the chromosomal length, centromere position, content and distribution of heterochromatin and satellite or secondary constriction characteristics. In pearl millet, differences have been reported in two varieties (AKP-3 and BIL-4) for arm ratios for some of the chromosomes (Pantulu, 1958; Venkateswarlu and Puntulu, 1968; Lobana and Gill, 1973). This difference has been attributed to the relative abundance of heterochromatin in BIL-4 and its limited occurrence in AKP-3. Pantulu *et al.* (1978) based on Giemsa–staining pattern suggested the possible occurrence of two types of heterochromatin at the centric regions. While all the pearl millet accessions studied were knobless, *P. purpureum* and *P. orientale* possessed several chromosomes with mostly terminal knobs (Pantulu, 1961, 1967; Sujatha, 1984).

At the interspecific level, the chromosome size shows considerable variation. The section *Gymnothrix* includes a species (*P. ramosum*) with very large chromosomes which, in fact, are regarded as the largest chromosomes in the genus itself (Jauhar, 1981). Even though pearl millet also has relatively larger chromosomes, the chromosomes of *P. ramosum* are about 5 per cent larger than those of pearl millet (Jauhar, 1981). This section also includes species with chromosomes of intermediate size (*P. massaicum*) and a species with small chromosomes

(*P. hohenackeri*). The sections *Eu-Pennisetum, Heteroshtachya* and *Brevivalvula* have species with generally small chromosomes (Jauhar, 1981).

Karyotype studies using different accessions of pearl millet revealed that most of the lines possessed 2–3 metacentric pairs, 2–3 sub-metacentric pairs and 1–2 acrocentric/sub-acrocentric pairs (Pantulu, 1961; Virmani and Gill, 1973; Tyagi 1975a; Sree Ramulu and Sree Rangaswamy1971; Sujatha, 1989; Vari and Bhowal 1982, 86). Deviation from this typical pattern has been reported by Chandola and Jain (1970) in different pearl millet varieties studied by them. They observed 3 telocentric pairs in some varieties of Sudan origin.

Majority of the pearl millet accessions studied possessed one pair of nucleolus organizing chromosomes which is the shortest pair (7th) with the NOR located in the short arm. However occurrence of two nucleolar-organizing chromosomes have also been reported (Krishnaswamy, 1962). Likewise, in some accessions the satellite was found to be located in the long arm of the fourth longest chromosome (Al-Fakhry *et al.*, 1964) while still in others it was located on the short arm of the largest (1st) chromosome (Gupta, 1966).

The role of B-chromosomes in contributing to overall diversity within the cultivated pearl millet has also been analyzed in considerable detail. B-chromosomes have originally been reported in African lines (Pantulu 1960, Burton and Powell, 1965, Powell and Burton, 1966). A great variation in the number of B's between different plants and also between different cells within a plant was found to be a common feature by several workers. The B-chromosomes described by Pantulu (1960) in the original African accessions possessed sub-terminal centromere with the short arm being totally heterochromatic while the long arm had a small proximal heterochromatic region followed by euchromatic region with 8 chromomeres. These were designated as standard type which were morphologically distinguishable from the derived types viz., deficient B's (lacking an euchromati segment with 2 chromomeres) and Iso B's (isochromosome for the long arm of deficient B; Subba rao and Pantulu, 1978). Euchromatic B's have also been known to occur in some accessions (Powell and Burton, 1966). These different types of B-chromosomes effected differently the plant morphology, mitotic cycle duration and meiotic behavior of A-chromosomes.

Diversity in basic Chromosome Numbers and Chromosome Ploidy

The most striking diversity in the chromosomal characters is the variation found in the number of sets of chromosomes as well as basic number among various species of this genus. Four basic numbers X = 5, 7, 8 and 9 have been reported in this genus and the studies made so far did not succeed in determining the original/ancestral/ phyletically basic number or in explaining the evolutionary trends of divergency in this feature-whether is it towards reduction or increase in basic chromosome number.

The basic chromosome number X = 5 is known to occur in only one species, *P. ramosum* whose genome exists in diploid condition only. Four species including the cultivated pearl millet viz., *P. glaucum, P. violaceum, P. schweinfurthii* and *P. purpureum* have X = 7 as their basic number. Of these, the first three occur in diploid state only while the fourth one, *P. purpureum*, has an allotetraploid genome (2n = 4x = 28). The basic chromosome number x = 8 finds its representation in only one species *P. mezianum* syn. *P. messaicum* where both diploid and autotetraploid cytotypes have been reported (Jauhar, 1975; Sujatha, 1989; Swaminathan and Nath, 1956; Rangaswamy, 1972). An aneuploid cytotype of this species with x = 34 (32 + 2Bs) has also been observed (Veyret, 1957).

Of the four basic chromosome numbers known, x = 9 finds greatest representation among the wild species which belong to the section *Eu-Pennisetum, Brevivalvula, Heterostachya* and *Gymnothrix*. Three species viz., *P. hohenackeri, P. flaccidum* and *P. orientale* show diploid genomes. Of these, with the exception of *P. hohenackeri*, autopolyploid cytotypes are reported for the other two species. Two other species with predominant polyploidy viz., *P. setaceum* with 9 bivalents + 9 univalents and *P. villosum* with 18 bivalents + 9 univalents in some of their cytotypes (Narayan, 1951; Raman *et al.*, 1962; Simpson and Bashaw, 1969; Rangaswamy, 1972) are supposed to involve allopolyploidy as well as autopolyploidy responsible for the variation (Table 3).

V. Biochemical Diversity

A) Phenolic Compounds

The diversity in the foliar phenolic compunds in twelve species of *Pennisetum* was analyzed by Subba Rao *et al.*, 1988. Out of the fifteen compounds identified, twelve viz., Quercetin, 3-glucoside, Rutin, Isochlorogenic acid, Neo-chlorogenic acid, Cis-*p*-coumaric acid, *p*-Hydroxybenzoic acid, Scopoletin, Trans-*p*-coumaric acid, Trans-sinapic acid, Vanillic acid and Cis-sinapic acid were found in seven species. (*P. americanum, P. purpureum, P. setaceum, P. divisum, P. orientale, P. squamulatum and P. polystachion*—Table 4). Cis-sinapic acid,

Table 3. Basic Chromosome numbers and cytotypes of *Pennisetum* species (Modified after Jauhar, 1981)

Sl.No.	Accession	x=	Cytotype	Section	Gene
1.	*P. ramosum* (Hoechst.) Schweing.	5	10	*Gymnothrix* (B-acc-Aerostigma)	T
2.	*P. glaucum* Syn *P. typhoides* (Burn.) Stapf *et* Hubb. syn *P. americanum* (L.) Leke.	7	14	*Penicillaria*	P
3.	*P. violaceum* Rich.	7	14	*Penicillaria*	P
4.	*P. schweinfurthii* Pilger	7	14	*Heterostachya*	
5.	*P. purpureum* Schumach.	7	28,56	*Penicillaria*	S
6.	*P. mezianum* Leke. (syn.*P. massaicum* Stapf)	8	16,32	*Gymnothrix* (B-acc-Aerostigma)	T
7.	*P. hohenackeri* Hoechst., ex Steud (*P. alopecuros* Steud.)	9	18	*Gymnothrix*	T
8.	*P. clandestinum* Hoechst. ex Chiov	9	36,54	T	
9.	*P. flaccidum* Griseb.	9	18,36,54	T	
10.	*P. nervosum* (Nees) Trin.	9	36,72	T	
11.	*P. notarisii* Dur.et Schiny. (syn. *P. pedicellatum* Trin.)	9	36,54	*Brevivalvula*	T
12.	*P. orientale* L.C.Rich	9	18,27, 36,45,54	*Heterostachya*	T
12A.	*P. frutescens* L.	9	63		
13.	*P. polystachyon* (L.) Schult. syn *P. longisetum* Poir., syn *P. triticoides* Poir.	9	36,54,63	*Brevivalvula*	T
14.	*P. setaceum* (Forsk.) Chiov. (syn *P. ruppellii* Steud)	9	27,54,	*Eu-Pennisetum*	T
15.	*P. setosum* (Swarty) L.C.Rich	9	54,78		
16.	*P. squamulatum* Fresen	9	54	*Heterostachya*	T
17.	*P. subangustum* (Schum.) Stapf et Hubb.	9	36,54	*Brevivalvula*	T
18.	*P. tempisquense* Pohl	9	72		T
19.	*P. villosum* R. Brown ex Fresen (syn. *P. longistylum* Vilm)	9	18,27,36, 145,54	*Eu-Pennisetum*	T
20.	*P. cenchroides* L.Rich.	9	36	*Eu-Pennisetum*	T
21.	*P. dubuim* (syn *P. macrorum* Trin)?	7	14–86		T

* P = Primary, S = Secondary, T = Tertiary

Table 4. Types of phenolic compounds in 12 species of *Pennisetum*
(Modified after Subba Rao *et al.*, 1988)

Compound	*P.americanum*	*P. violaceum*	*P. purpureum*	*P. mezianum*	*P. hohenackeri*	*P. setaceum*	*P. divisum*	*P. orientale*	*P. clandestinum*	*P. villosum*	*P. squamulatum*	*P. polystachion*
Quercetin,3-glucoside	**	**	*	*	*	*	**	**	**	*	**	**
Rutin	**	**	*	*	*	*	**	*	**	*	*	*
Isochlorogenic acid	*	–	*	*	–	*	*	*	–	*	*	*
Chlorogenic acid	**	**	**	**	**	**	**	**	**	**	**	**
Neo–chlorogenic acid	***	***	***	**	***	***	**	***	***	**	***	**
Cis-p-coumaric acid	**	–	*	*	**	**	***	**	***	**	**	*
p-Hydroxybenzoic acid	**	***	*	**	***	***	**	*	**	–	**	**
Scopoletin	**	***	***	**	***	***	***	***	**	***	**	**
Trans-p-coumaric acid	**	**	*	*	**	**	*	**	**	**	**	*
Trans-sinapic acid	***	**	**	**	**	**	**	***	**	**	**	**
Vanillic acid	***	***	***	***	***	***	***	***	***	***	**	**
Cis-sinapic acid	***	*	*	–	**	**	*	***	*	**	*	*
Phloretic acid	–	**	–	–	–	–	–	–	–	–	–	–
Homo protocatechuic acid	–	–	–	–	*	–	–	–	–	–	–	–
Protocatechuic acid	–	–	–	–	*	–	–	–	–	–	–	–

*** Strongly present; **Distinctly present; * Faintly present;–Absent

Iso-chlorogenic acid and *p*-Hydroxybenzoic acid were lacking in *P. mezianum*, *P. clandestinum* and *P. villosum* respectively. Two additional compounds (Homo protocatechuic acid and Protocatechuic acid) were noticed in *P. hohenackeri* while Phloretic acid was specifically present in *P. violaceum* only.

Bate-Smith (1968) reported occurrence of Caffeic acid, *p*-coumaric acid and ferulic acid in *P. orientale*, where as Subba Rao *et al.*, (1988) noticed both Cis and Trans forms of *p*-coumaric acid and two derivatives of Caffiec acid (Iso-chlorogenic and Chlorogenic acids). In addition to these, Sinapic acid was also detected by the later workers. These differences in the phenolic composition observed between the two reports were attributed to the differences in the material investigate since several chromosomal types with 2n = 18 to 45 are known to exists in these species.

The percentage similarities between the species ranged from 69.20 in *P. violaceum* vs *P. mezianum* to 100 for several pairs of species compared. The two species *P. americanum* and *P. purpureum* showed 100% similarity while *P. violaceum* was showing a similarity of 76.9% with both the species. Both *P. americanum* and *P. purpureum* showed the same percentage similarities with *P. mezianum* as well as with *P. hohenackeri* while *P. violaceum* was showing less similarity with these two species. These observations are of interest when the relationships of the 3 species, *P. americanum*, *P. violaceum* and *P. purpureum*, are considered. *P. purpureum* is an allotetraploid with one of its genomes homologous to *P. americanum*. In spite of this, data on the phenolic compounds showed 100% similarity between these two species. *P. violaceum* was considered to be the probable progenitor of the cultivar *P. americanum* based on hybridization experiments (Brunken, 1977). But the relation between these two genomes is only 76%. Based on these percentage similarities, *P. violaceum* is equidistant from both *P. americanum* and *P. purpureum*.

B) Diversity in Seed Nutritive Value

In view of the nutritive value of the grain of the cultivated species, considerable information on the genetic diversity contributing to this character is also available in pearl millet. Pearl millet endosperm contains a mixture of starch granules, studded with protein bodies (Rooney and Mc Donough, 1987). The major nutritional constituents are carbohydrates, proteins, lipids and ash. The grains are also a rich source of minerals, trace elements, vitamins and specific free amino acids. Information available on the quantitative variation of these chemical constituents in different races/varieties of pearl millet is summarized in Table 5.

C) Protein Diversity

The conserved genetic status of seed storage proteins and their relative ease of use in analyzing large numbers of samples make this biochemical genetic marker a useful trait in studying some facets of species relationships. (Johnson and Hall, 1965). There were no differences in albumin subunit composition in the *monodii* sub species of the cultivated pearl millet. Similar profiles were observed for high and low molecular weight for representative accessions of the tertiary gene pool—*P. ramosum, P. schweinfurthii* and *P. squamulatum* that had basic chromosome numbers 5, 7 and 9 respectively. The globulin subunit

Table 5. Intraspecific variation in Protein, Amino acid and Fatty acid composition in different land races of pearl millet.
(Modified after Faujdar Singh and Nainawatee, 1999)

Item	Range	Item	Range
1. Total Proteins (%)	6.04–24.25	**5. Fatty acids (%)**	
2. Protein in different fractions (g/100g)		Palmitic acid	19.30–23.40
		Palmitoleic acid	0.37–0.83
Albumin	1.22–3.17	Stearic acid	2.56–4.64
Globulin	0.46–1.55	Oleic acid	21.20–27.10
Prolamine	2.33–5.89	Linoeic acid	40.40–51.70
Glutelin	2.29–5.60	Linolenic acid	3.30–4.70
		Arachidic acid	0.30–0.70
3. Amino acids (g/16g N)		**6. Chemical composition**	
		A. Major constituents (%)	
Lysine	1.59–3.80	Starch	62.8–70.5
Leucine	8.62–14.80	Amylose	21.9–28.8
Isoleucine	3.70–6.34	Soluble sugars	1.4–2.6
Methionine	1.02–2.12	Reducing sugars	0.1–0.26
Cysteine	1.43–3.69	Crude fibre	1.1–1.8
Threonine	3.17–5.66	Ash	1.1–2.5
Valine	4.38–7.67	*B. Minerals & Trace elements (mg/100g of grain)*	
Phenylalanine	3.82–4.94		
Tyrosine	6.54–10.81		
Tryptophan	1.66–1.83	Potassium	294.0–460.0
Histidine	2.21–2.48	Phosphorus	185.0–363.0
Arginine	4.82–5.75	Magnesium	46.0–128.0
4. Vitamines (mg/100g grain)		Calcium	13.0–52.0
		Iron	4.0–58.1
Carotene	85–445	Zinc	1.0–6.6
Thiamine	0.2–0.33	Copper	0.6–21.2
Riboflavin	0.2–0.3	Manganese	0.2–1.8
Niacin	2.3–3.5		

composition revealed that *P. schweinfurthii* was very different in being characterized by the presence of major sub units of low molecular weight and the absence of some of the prevalent globulin subunits of medium molecular weight associated with other species.

Prolamine analysis with the Urea-PAGE system was found to be more suitable than were the albumin and globulins in studying the differences within and between species of the *Pennisetum* gene pool. The prolamine composition within the range of the primary gene pool which spans nearly the entire Sahalian zone showed that majority of the accessions were characterized by the prevalence of at least four bands designated A1, A2, B1 and B2. Similar patterns occurred in pearl millet land races including those from Zimbabwe and India. Prolamine polymorphism levels were much higher in the wild *monodii* subspecies than in cultivated pearl millet. The accessions from Chad showed an ecotypic distribution of prolamine types into three groups. Two populations from eastern Chad possessed bands distinct from those of a central Chad population as well as from two other populations of western Chad. Prolamine types of the latter population, which were more common among the wild subspecies, could be found as far west as Mauritiana.

Prolamine analysis provided an analytical tool for the evaluation and selection of germplasm in *P. purpureum*. All species in the tertiary gene pool revealed differences in prolamine composition. Geographical diversity within species was evident in accessions of *P. ramosum* from Chad and Ethiopia, in *P. orientale* from Pakistan and Australia and in *P. nervosum* from Mexico and Puerto Rico.

D) Enzyme Diversity

In addition to the study of genetic diversity in terms of protein patterns, another level of biochemical diversity analysis has also been conducted using isoenzyme patterns for alcohol dehydrogenase (ADH) in association with early Vs. late maturation character. In the West African populations, the early and late maturing cultivars formed two distinct enzyme groups. With reference to Esterase (est), Catalase (cat), Phospho glucoisomerase (PGI), Phospho glucomutase (Pgm), 6-phospho glucose dehydrogenase (Pgd), glutamate–oxaloacetate transaminase (Got) and Malate dehydrogenase (Mdh). (Tostain *et al.*, 1987; Tostain and Marchais, 1989).

A total of 199 samples of landraces of pearl millet collected from West Africa (106), East Africa (32), Southern Africa (30) and India (31) were analyzed for the enzyme polymorphism. The analysis revealed Indian millets to be very distant from the southern African group but closer to Niger and the late maturing group of West Africa. Enzyme diversity was maximum in the early west group and decreased in the following order—Early East > Late West and East > India and Southern Africa. For each region, the late group was always less diverse than the early one. In regions where pearl millet is a major crop (the early West, Early East and Indian groups) intra-population diversity formed 80–90 per cent of the total diversity whereas in regions where the crop is of minor importance (mainly late maturing cultivars), the intra-population diversity was below 70% of the total diversity observed. Such high intra population diversity was attributed to the high levels of natural allogamy in pearl millet coupled with genetic drift. The Nei's diversity distance values indicated that the inter-group diversity was only 6.7% of the total diversity.

The genes that are quasi-monomorphic in the early west group remained so in the other groups also. While the rare alleles found in west African group did not increase their frequency in the other groups, some new but rare alleles (MdhC, AdhA, EstA nd GotA) were found outside of West Africa. The order of genes according to decreasing diversity was EstA > Adh > PgmA > catA > PgiA > PgdA > MdhA > MdhC > PgdB > GotB > GotA > Mdh and this order prevailed in all the groups. Among the alleles, which generally discriminated the groups, the most frequent allele in the early West group showed a tendency towards fixation in the other groups. Based on these inferences, viz., (i) the occurrence of all pearl millet enzyme diversity in samples from a unique geographical region (Mali-Niger-East Africa), (ii) the low inter-regional diversity, (iii) the early–late disjunction in west African populations observed even at the level of cultivars from the same village, irrespective of ecological zone or country, and (iv) lack of significant early-late differences outside West Africa, Tostain and Marchais (1989) suggest that environment had little effect on the organization of pearl millet enzyme diversity and attributed the existing diversity to migrations, foundations and genetic drift from a unique gene source.

A comparative study of isozyme diversity at interspecific level in *Pennisetum* was carried out by Sujatha (1984), Saideswara Rao *et al.*, (1986), Lagudah and Hanna (1989) and Hussain *et al.*, (1990). Of all the enzymes studied, the extent of polymorphism was relatively high for Esterase (Est), Aminopeptidase(AMP), Shikimate dehydrogenase (SKDH) and was least for b-phosphogluconate dehydrogenase (PGD). The taxonomic relationships (based on similarity indices) arrived at in terms of the isozyme variations is largely, in agreement with that arrived at on the basis of cytological evidence. SKDH and Est revealed maximum diversity among the tertiary gene pool species (Lagudah and Hanna, 1989). Unlike in the case of intraspecific enzyme diversity, the information at the interspecific level is less conclusive, for example, species of the *Brevivalvula* section are shown to be closely related on the basis of monomorphic PGD zymograms and esterases but seem highly divergent based on AMP.

VI. Diversity at Molecular Level

Diversity in terms of DNA sequence has also been reported at the interspecific level in the genus *Pennisetum*. In the chloroplast DNA, the inverted repeat regions showed least diversity between the cultivated pearl millet and tertiary gene pool species whereas the restriction sites of single copy regions were relatively more variable (Clegg *et al.*, 1984). Studies of mitochondrial DNA revealed considerable diversity between pearl millet and seven wild species. Using six mitochondrial gene sequences as probes, it was found that *P. purpureum* and *P. squamulatum* show least diversity and in this aspect were more closer to pearl millet than to other species in the tertiary gene pool (Chowdhury and Smith, 1988).

Gepts and Clegg (1989) analyzed the RFLPs among chloroplast DNA (cDNA), nuclear ribosomal RNA (rRNA) and alcohol dehydrogenase (ADH) sequences in a group of 25 wild and 54 cultivated accessions of pearl millet to find out which type of DNA sequence among these would be most informative about the distribution of genetic diversity. They observed no polymorphism for cDNA sequences but the nuclear rRNA genes exhibited greater variation among the wild accessions. Though the Adh_1 genes showed variability between wild and cultivated types, they did not provide additional information on the area of domestication, since the most frequent Adh_1 pattern among cultivars is also the most frequent and geographically widespread pattern among wild types.

Vii. Significance of Genetic Diversity Anlysis in *Pennisetum*

Besides the use of genetic diversity in crop improvement programs, analysis of the diversity using different types of markers has helped to understand the probable mechanism and pathways of domestication and evolution within the genus. The use of combination of taxonomic characters which are likely to be neutral to natural selection and physiological characters which are dependent on environment, led to the suggestion that Indian pearl millets were distinct from, and less diverse than, African populations (Upadhyay and Murthy, 1970). Porteres (1976) basing on botanical (morphological) characters alone, found that African pearl millet clustered essentially into three groups according to geographical origin—West, East and South with the Indian land races close to the South African group. In addition, this analysis also enabled the realization of an association between the observed morphological groups in pearl millet and the main agrarian civilizations viz., Paleo-nigritic, Nilo-cushitic and Bantu. This morphological diversity has also led to the suggestion that pearl millet domestication occurred most probably in the West African Sahel (Brunken *et al.*, 1977) where the cross-fertile wild species (and probable progenitor) *P. violaceum* is also currently found. The analysis of enzyme diversity has provided more conclusive information on the mechanism of diversity and pattern of migrations in this genus. The strong divergence between Indian and South African populations and their relative proximity with West African populations (which are most diverse ones) have been interpreted as being suggestive of independent origin of Indian and South African races within West Africa. The similarity between samples from India and Niger and other areas (Sudan, Uganda, Kenya and Somalia) which group with Indian types would seem to indicate a path of migration from Niger or the Central Sahel to India via the Arabian coast, Iran and Pakistan, (Tostain and Marchais, 1989). This interpretation of the results based on enzyme analysis together with information on linguistic (Jarrige, 1977), archaeological (Davis, 1968; Cleuzious and Constantini, 1982) and domestication of Sorghum (Harlan and Stember, 1976; Ollitrault, 1987) suggest that Sorghum and pearl millet might have migrated together from their African center(s) of origin. The sequential events involved in the domestication and evolution of pearl millet suggested include: multiple site domestication in the African Sahel leading to—(i) creation of early maturing cultivars and their migration to India plus a southward migration into the Sudanian zone, (ii) creation of late maturing varieties—continuation of the migration simultaneously westwards and east and southwards to southern Africa (Tostain and Marchais, 1989).

REFERENCES

Al-Fakhry, A.K., Sarvella, P. and Grogan, C.O. 1964. Somatic chromosomes of Babala. J. Hered. **55:** 57–60.

Andre Savery, M. and Prasad, M.N. 1995. Genetic diversity in pearl millet, *Pennisetum typhoides*. Madras Agri. J. **82:** 8–11.

Appadurai, R., Raveendran, T.S. and Nagarajan, C. 1982. A new male sterility system in pearl millet. Indian J. Agric. Sci. **52:** 832–834.

Appa Rao, S. and de Wet, J.M.J. 1999. Taxonomy and evolution. In: Pearl millet breeding. Ed. Khairwal, I.S., Rai, K.N., Andrews, D.J. and Harinarayana, G. Oxford and IBH Co. Pvt. Ltd., New Delhi. p. 29–48.

Athwal, D.S. 1961. Recent development in the breeding and improvement of bajra (pearlmillet) in Punjab. Madras Agric. J. **48:** 18–19 (Abstr.).

Avdulov, N.P. 1931. Karyosystematische untersuchung der Familie Gramineen. Bull. Appl. Bot. Pl. Breed. Suppl. **4:** 1–428.

Bate-Smith, E.C. 1968. The phenolic constituents of plants and their taxonomic significance II. Monocotyledons. J. Linn. Soc. Bot. **61:** 325–356.

Belliard, J., Nguyen, E.V. and Pernes, J. 1980. Analyse des relations genetiques entre formes spontanees et cultivees chezle Mil a chandelle. I.—Etude des parents et des hybrids de premiere generations (F1) entre un ecotype de *Pennisetum mollissimum* Hochst (Forme spontanee) et differentes formes cultivees de *Pennisetum americanum* (L.) Leeke. Annuale Amelioration des Plantes **30:** 229–251.

Bilquez, A.F. and Lecomte, J. 1969. Relations entremil sauvage et mils cultives: etude de I" hybridide *Pennisetum typhoides* Stapf. and Hubb. X *P. violaceum* (Lam.) L. Rich. L' Agronomie Tropicale **24:** 249–257.

Branek-loox, P.J., Andrews, D.J., Bidinger, F.R. and Frey, K.J. 1984. A rapid method of evaluating growth rate in pearl millet and its weedy and wild relatives. Crop Science **24:** 1187–1191.

Brunken, J.N. 1977. A systematic study of *Pennisetum* Sect. *Pennisetum* (Gramineae). Ame. J. Bot. **64:** 161–176.

Brunken, J., de Wet, J.M.J. and Harlan, J.R. 1977. The morphology and domestication of pearlmillet. Econ. Bot. **31:** 163–174.

Burton, G.W. 1958. Cytoplasmic male sterility in pearlmillet (*P. glaucum*) Agron. J. **50:** 230–235.

Burton, G.W. 1965a. Pearlmillet 'Tift 23 A' released. Crops Soils **17:** 19.

Burton, G.W. 1965b. Cytoplasmic male sterile pearlmillet 'Tift 18 A'. Crops Soils **18:** 19.

Burton, G.W. and Powell, J.B. 1965. Six chlorophyll-deficient seedlings in pearlmillet, *Pennisetum typhoides,* and a suggested system for their nomenclature. Crop Sci. **5:** 1–3.

Chandola, R.P. and Jain, S.N. 1970. Karyomorphological studies in *Pennisetum typhoides* Stapf et Hubbard. Curr. Sci. **19:** 92–93.

Chatterji, A.K. and Pillai, G.K. 1970. Apomixis in *Pennisetum pedicellatum.* Trin Sci: Cult. **36:** 669–670.

Chatterji, A.K. and Timothy, D.H. 1969a. Microsporogenesis and Embryogenesis in *Pennisetum flaccidum* Griseb. Crop Sci. **9:** 219–222.

Chowdhury, M.K.V. and Smith, R.L. 1988. Mitochondrial DNA variation in pearlmillet and related species. TAG **76:** 25–32

Clark, J.D. 1962. The spread of food production in sub-Saharan Africa. J.Afr. Hist. **3:** 211–278.

Clegg, M.T., Rawson, J.R.Y. and Thomas, K. 1984. Chloroplast DNA variation in pearlmillet and related species. Genetics **106:** 449–461.

Cleuzious, S. and Constantini, L. 1982. A L'origine des Oasis Rocher che. **137:** 1180–1182.

Davies, O. 1968. The origin of agriculture in West Africa. Curr. Anthro. **9:** 479–482.

D'Curz, R. and Reddy, P.S. 1968. Apomixis in *Pennisetum massaicum* Stapf. Sci.Cult. **34:** 255–257.

Gepts, P. and Clegg, M.T. 1989. Genetic diversity in pearl millet (*Pennisetum glaucum* (L.) R.Br.) at the DNA sequence level. J.Hered. **80:** 203–208.

Gupta, V.P. and Athwal, D.S. 1966a. Genetic variability correlation and selection indices of grain characters in pearlmillet. J. Res. P. AU **3:** 111–117.

Gupta, V.P. and Athwal, D.S. 1966b. Genetic variability correlation and selection indices for green fodder characteristics in pearlmillet. J. Res. P. AU **3:** 379–384

Hanna, W.W. 1979. Intesrspecific hybrids between pearlmillet and fountain grass. J. Hered. **70:** 425–427.

Hanna, W.W. 1987. Utilization of wild relatives of pearlmillet. Pages 33–42 In: Proceedings international Pearl Millet Workshop. (J.R. Witcobe ad S.R. Beckerman, eds), ICRISAT, Patancheru, India.

Hanna, W.W. 1990. Transfer of germplasm from the secondary to the primary gene pool in *Pennisetum*. Theoretical and Applied Genetics **80:** 200–204.

Hanna, W.W. and Dujardin, M. 1982. Apomictic interspecific hybrids between pearl millet and *Pennisetum orientale* L.C. Rich. Crop Sci. **22:** 857–859.

Hanna, W.W., Powell, J.B. and Burton, G.W. 1976. Relationship to Polyembryony, frequency, morphology, reproductive behaviour and cytology of autotetraploids in *Pennisetum americanum*. Can. J. Genet. Cytol. **18:** 529–536.

Harlan, J.R. 1971. Agricultural origins: Centers and non-centers. Science **14:** 468–474.

Harlan, J.R. 1975. Crops and Man. American Society of Agronomy, Madison, USA

Harlan, J.R. and de Wet, J.M.J. 1971. Towards a rational taxonomy of cultivated plants. Taxonomy **20:** 509–517.

Harlan, J.R. and Stember, A.B.L. 1976. The races of sorghum in Africa. In Harlan, J.R., De Wet, J.M.J., Stemler, A.B.L. (eds) Origins of African plant domestication. Mouton, The Hague Paris, pp. 465–478.

Hrishi, N.J. 1952. Studies on the cytogenetics of some species of *Pennisetum* and their comparative morphology and anatomy. Genetica **26:** 280–356.

Hussain, S.N., Saideswara Rao, Y., Manga, V. and Subba Rao, M.V. 1990. Chaemotaxonomic studies on the genus *Pennisetum* L. Rich by isozyme analysis. Beitrage ur Biologie der Pflanzenzuchtung. **65:** 211–222.

Jarrige, J.F. 1977. La civilisation de I' Indus. Recherche **76:** 245–252.

Jauhar, P.P. 1975. Genetic control of chromosome pairing in polyploid fescues: its phylogenetic and breeding implications. Rept Welsh Pl Breed Stn. For 1974, pp. 114–127.

Jauhar, P.P. 1981. The eternal controversy of the latin name of pearlmillet. Indian J. Bot. **4(1):** 1–4.

Johnson, B.L. and Hall, O. 1965. Analysis of phylogenetic affinities in the triticeae by protein electrophoresis. Ann. J. Bot. **52:** 506–513.

Kajjari, N.B. and Patil, J.A. 1956. A male sterile bajra *Pennisetum typhoides* (Brun). S. & H. Indian J. Genet. Pl. Breed. **16:** 146–147.

Khairwal, I.S. and Sultan Singh. 1999. Quantitative and qualitative traits. In: Pearlmillet Breeding. Eds.Khairwal, I.S., Rai,K.N., Andrews, D.J. and Harinarayana, G. Oxford and IBH Co. Pvt. Ltd., New Delhi. pp. 119–155.

Krishna Rao, M. and Koduru, P. R.K. 1978. Inheritance of genetic male sterility in *Pennisetum americanum* (L.) Leeke. Euphytica **27:** 777–783.

Krishnaswamy, N. 1962. Bajra. *Pennisetum typhoides* S. & H. ICAR, New Delhi

Lagudah, E.S. and Hanna, W.W.1989. Species relationship in the *Pennisetum* gene pool. Enzyme polymorphism. TAG **78:** 801–808.

Lobana, K.S. and Gill, B.S. 1973. Pachytene chromosomes of *Pennisetum typhoides*. Cytologia **38:** 713–717.

Manga, V., Sujatha, D.M. and Subba Rao, M.V. 1991. Pollen morphology in cultivated and wild species of genus *Pennisetum*. Special Publication No. 16. In: A decade of Scanning electron microscopy in Geological Survey of India.

Mangat, B.K., Maiti, R.K. and Khairwal, I.S. 1999. Pearl millet Biology. In: Pearlmillet Breeding. Eds. Khairwal, I.S., Rai, K.N., Andrews, D.J. and Harinarayana, G. Oxford and IBH Co. Pvt. Ltd., New Delhi. pp. 1–27

Marchais, L. and Tostain, S. 1983. New male sterile cytoplasms in pearl millet. XV Int. Cong. Genet. New Delhi. 174pp (Abstr.).

Marchais, L. and Tostain, S. 1985. Genetic divergence between wild and cultivated perlmillet (*Pennisetum typhoides*) II. Characters of domestication. Zeitschrift fur Pflanzenzuchtung **95:** 245–261.

Mukherji, P., Agarwal, R.K. and Singh, R.M. 1981. Component and combining ability analysis in a 8 × 8 diallel of pearlmillet. Madras agric. J. **68:** 436–443.

Muddock, G.P. 1959. Africa- Its people and their cultural history, Mc Graw Hill, New York.

Nair, V.G. and Pillai, P.K. 1969. Bulbils in *Pennisetum polystachyon*. Agric. Res. J. Kerala **6:** 54.

Narayan, K.N. 1951. Cytogenetic studies of apomixis in *Pennisetum*. Ph.D. Thesis. University of Calif., Berkely.

Narayan, K.N. 1954. Meiosis in the pollen mother cells of *Pennisetum ramosum* (2n=10) Curr. Sci. **23:** 407–408.

Narayan, K.N. 1955. Cytogenetic studies of apomixis in *Pennisetum*. Proc. Indian Acad. Sci. B. **41:** 196–208.

Narayan, K.N. 1962. Apomixis in some species of *Pennisetum* and in *Panicum antidotale*. p. 55–61, In "Plant Embryology" Proc. Sum. Univ. Delhi, Council of Scientific and Industrial Research, New Delhi.

Nunez, O. 1952. Investigaciones cariosistematicas en las Gramineas Argentinas de la tribus "Paniceae" (Karyosystematic investigations on the Argentinean grasses of the trine Paniceae) Rev. Fac. Agron, La Plata **28:** 229–256 (Spanish).

Ollitrault, P. 1987. Evaluation genetique des sorghos cultives (Sorghum bicolor L. Moench) par 1' analyse conjointe des diversites enzymatiques et morphophysiologiques Relations Los Angels, London, pp. 272–280.

Pantulu, J.V. 1958. A case of chromosomal interchange in pearl millet. Curr. Sci. **27:** 497–498.

Pantulu, J.V. 1960. Accessory chromosomes in *Pennisetum typhoides*. Curr. Sci. **29:** 28–29.

Pantulu, J.V. 1961. Cytological studies in the genus *Pennisetum* with some cytological observations in the genus *Cassia*. Ph.D. Thesis, Andhra University, Waltair (India).

Pantulu, J.V. 1967. Pachytene pairing and meiosis in the F1 hybrid of *Pennisetum typoides* and *P. purpureum*. Cytologia **32:** 532–541.

Pantulu, J.V., Krishna Rao, M., Ganapathi, K. and Subba Rao, B. 1978. Giemsa staining of heterochromatin of A– and B- chromosomes of *Pennisetum typoides* (Burm.) S. & H. Indian J. Bot. **1:** 101–104.

Patil, B.D., Hardas, M.W., O'connor, F.K. and Vohra, S.K. 1962. Polyploidy in *Pennisetum orientale* Rich. Curr. Sci. **31:** 161–162.

Porteres, R. 1962. Berceaux agricoles primaries surle continent Africain. J.Afr. Hist, **3:** 195–210.

Porteres, R. 1976. African cereal: *Eleusine*, fonio black fonio, teff, *Brachiaria*, *Paspalum*, *Pennisetum* and African rice. Pages 409–452. In: Origins of African Plant Domestication. (J.R. Harlan, J.M.J. de Wet and ABL Stemler, eds). Mouton Press, The Hague, The Netherlands.

Powell, J.B. and Burton, G.W. 1966. Nucleolus-organizing accessory chromosomes in pearl millet *Pennisetum typhoides*. Crop Sci. **6:** 131–134 .

Powell, J.B. and Burton, G.W. 1968. Polyembryony in pearl millet, *Pennisetum typhoides*. Crop Sci. **8:** 771–773.

Powell, J.B., Hanna, W.W. and Burton, G.W. 1975. Origin, cytology, and reproductive characteristics of haploids in pearl millet. Crop Sci. **15:** 389–392.

Purseglove, J.W. 1976. Millets. Pages 91–93 In: Evolution of crop plants. (N.W. Simmonds, ed.) Longman, London, UK.

Raman, V.S., Krishnaswamy, D. and Nair, M.K. 1962. Studies on the progeny of auto-tetraploid *Pennisetum typhoides*, Stapf & Hubb. J. Indian bot. Soc. **41:** 268–270.

Rangaswamy, S.R.S. 1972. Cytological studies on diploid and polyploid taxa of the genus *Pennisetum* Rich. Genetica **43:** 257–273

Rooney, L.W. and Mc Donough, C.M. 1987. Food quality and consumers acceptance of pearl millet. Pages 43–61 In: Proceedings International Pearl millet Workshop. (J.R. Witcombe and S.R. Beckerman, eds.) ICRISAT, Patancheru, India

Saideswara Rao, Y., Sujatha, D.M., Manga, V. and Subba Rao, M.V. 1986. Esterase variability in five species of the genus *Pennisetum* L. Rich. (Gramineae). Geobios **13:** 108–111.

Shantamma, C. and Narayan, K.N. 1977a. Studies in Poaceae (Graminee). J. Mysore Univ. B, **27:** 302–305.

Shantamma, C. and Narayan, K.N. 1977b. Formation of nucellar embryos with total absence of embryo sacs in two species of Gramineae. Ann. Bot. **41:** 469–470.

Simpson, C.E. and Bashaw, E.C. 1969. Cytology and reproductive characteristics in *Pennisetum setaceum*. Amer. J. Bot. **56:** 31–36.

Sree Ramulu, K. and Sree Rangaswamy, S.R. 1971. Karyomorphological studies in *Pennisetum*. Cytologia **36:** 480–486.

Stapf, O. and Hubbard, C.E. 1934. Gramineae Vol. 9. In: Flora of Tropical Africa Prain, D. (Editor). Kent, England Reeve and Co. Ltd.

Stebbins, G.L. 1971. Chromosomal evolution in higher plants. Edward Arnold Ltd. London.

Subba Rao, M.V. and Pantulu, J.V. 1978. The effects of derived B-chromosomes on meiosis in pearl millet *Pennisetum typhoides*. Chromosoma (Berl.) **69:** 121–130

Subba Rao, M.V., Sujatha, D.M., Saideswara Rao, Y. and Manga, V. 1988. Chemotaxonomic characters in twelve species of the genus *Pennisetum* (Poaceae). Proc. Indian Acad. Sci. (Plant Sci.) **98.2:** 111–120

Sujatha, D.M. 1984. Studies on some biosystematic aspects of twelve species of the genus *Pennisetum* L. Rich. (Poaceae) Ph.D. Thesis. Andhra University, Visakhapatnam, India

Sujatha, D.M., Manga, V., Subba Rao, M.V. and Murthy, J.S.R. 1989. Meiotic studies in some species of *Pennisetum*. Cytologia **54:** 641–652

Swaminathan, M.S. and Nath, J. 1956. Two new basic chromosome numbers in the genus *Pennisetum*. Nature **178:** 1241–1242

Tostain, S. and Marchais, L. 1989. Enzyme diversity in pearlmillet (*Pennisetum glaucum*) 2. Africa and India. Theoretical and Applied Genetics **77:** 634–640

Tostain, S., Raindey, M.F. and Marchais, L. 1987. Enzyme diversity in pearlmillet (*Pennisetum glaucum*) 1. West Africa. Theoretical and Applied Genetics **74:** 188–193.

Tyagi, B.R. 1975a. Karyomorphology of somatic chromosomes in pearlmilet. Proc. Indian Natn. Sci. Acad Section B **41:** 462–465.

Upadhyay, M.K. and Murty, B.R. 1970. Genetic divergence in relation to geographical distribution in pearl millet. Indian Journal of Genetics and Plant Breeding **30:** 704–715.

Upadhyay, M.K. and Murty, B.R. 1971. Genetic diversity and combining ability in pearlmillet. Indian Journal of Genetics and Plant Breeding **31:** 63–71.

Vari, A.K. and Bhowal, J.G. 1982. Aneuploids of *Pennisetum typhoides*. Cytologia **47:** 803–808.

Vari, A.K. and Bhowal, J.G. 1986. Studies on the trisomics of *Pennisetum americanum* (L.) Leeke. Morphological and cytological behaviour of primary trisomics. Cytologia **51:** 679–692.

Vavilov, N.I. 1950. The origin, variation, immunity and breeding of cultivated plants. Chronica Botanica **13:** 1–366.

Venkateswarlu, J. and Pantulu, J.V. 1968. Morphology of Pachytene chromosomes in pearl millet. Heridity **59:** 69–70.

Veyret, Y. 1957. Les chromosomes somatiques chez quelques especes de *Pennisetum*. (The somatic chromosomes in some species of *Pennisetum*). Agron Trop. **12:** 595–598 (French).

Virmani, S.S. and Gill, B.S. 1972. Somatic chromosomes of *Pennisetum typhoides* (Burm.) S.H. Cytologia **37:** 257–260.

CHAPTER 5

AQUATIC FLOWERING PLANTS DIVERSITY OF COASTAL PLAINS OF KONKAN, GOA AND KARNATAKA

S.P. Gaikwad, M.M. Sardesai* and S.R. Yadav***

Introduction

World civilizations and cultures have been centered on fresh water habitats. The world's rivers, lakes, streams and wetlands provide most of the world's water for drinking, agriculture, sanitation and industry as well as huge quantities of fish and aquatic organisms. Fresh waters are also home for great diversity of fishes, amphibians, invertebrates, aquatic plants and microorganisms. Freshwater lakes are classical example of "Habitat Island" and ancient lakes tend to possess high level of endemism.

Wetlands play important role in any ecosystem and hydrology of area. Regulatory functions such as downstream flood control, ground water recharge and sediment or nutrients retention are examples of ecological functions performed by wetlands that may indirectly benefit economy. They play important role in shoreline, bank stabilization, recreation, tourism, water transport and provide habitat for migratory birds, fisheries, water plants, animals and microbes.

Aquatic macrophytes have many actual or potential uses as compost, mulches, fertilizers, human food, feeds for livestock, source for fibers for paper making and source of various chemical substances

* Department of Botany, Shivaji University, Kolhapur-416 004.
** Department of Botany, Delhi University, Delhi-110 007.

(Sculthrope, 1967; Little, 1968; Boyd, 1972). They are also used for the purification of pollutant effluents and water bodies (Byod, 1970c; Yount and Crossman, 1970; McNabb *et al.*, 1971; Seidel, 1971). Rice, an emergent aquatic is a major agronomic species. In addition to rice, number of aquatic plants which form a source of food and vegetable (Byod, 1972) and many of them are grown in small ponds for ornamental purposes. The sale of plants for aquarium is of growing economic importance.

Excessive growth of aquatic plants and problem created by aquatic plants have attracted attention of several workers, however, number of aquatic plants play important role beneficial to man. In addition to their importance as a food, compost and fertilizers, aquatic plants play important role in removal of nutrients, heavy metals and purification of water (Mitchell, 1974; Rai *et al.*, 1995). Aquatic plants can possibly be employed to remove the nutrients and heavy metals from wastewater effluents. Plants such as *Phragmites communis, Scripus lacustris, Eichhornia crassipes, Spirodella polyrrhiza, Elodea canadensis, Egeria densa, Hydrilla* species, *Ceratophyllum demersum, Bacopa monnieri* and *Nymphoides cristata* have been found suitable for removal of different metals (Rai *et al.*, 1995). The plants remove the metals by surface adsorption and/or absorption and incorporate them into their own system or store them in a bound form. Effluents ameliorated by these plants thus cause less damage to aquatic environment. For developing countries like India, such technology would be extremely important where shallow ponds and wetlands are unsuited for conventional fish farming and agriculture (Mohan Ram, 1978).

Inspite of number of important functions performed by wetlands, they are continuously destroyed all over the world. According to Dugen (1990) the U. S. has lost some 54% of its wetlands, New Zealand has lost over 90 % of its natural wetlands and many parts of Europe have lost nearly all natural wetlands and tropical countries like Chad, Cameroon, Niger, Bangladesh, India, Thailand and Vietnam have lost over 80% of their fresh water wetlands (McNeely *et al.*, 1990; McNeely 1992). Biodiversity crises all over the world in 20th century resulted in number of important publications on worldwide loss of biodiversity. Rio Earth Summit (1992) emphasized the need to conserve biodiversity of Globe especially the tropics. With growing awareness, the importance of wetlands has been realized throughout the world (Gopal, 1990; Anonymus, 1992; Chatrath, 1992). The International Hydrological Decade (IHD) 1965–74 was launched by the General Conference of UNESCO promotes international co-operation in research, studies and training of specialists and technicians in scientific hydrology. Infact the RAMSAR convention is called convention of wetlands of International importance. The Indian subcontinent also called the aquatic subcontinent comprises 1,28,000 sq. km. of water surface of which 77,000 sq. km. occur in India. The subcontinent hosts approximately 470 aquatic species of flowering plants, a half of the worlds known aquatic angiosperms and the number of endemic species also being high (13%), second to tropical South America (Gopal, 1990).

India has already lost over 80% of its fresh water wetlands. There is an urgent need to conserve the wetlands and their biodiversity. The meaningful conservation involves thorough understanding of the flora of a particular region. The inventory must be completed at faster rate, least the developmental activities are sure to eliminate some of the species even before we understand fully. Considering role-played by the wetlands, their fauna and flora, it becomes important to make detailed inventory of them. Similarly, numbers of indigenous aquatic flowering plants of our country are demising their populations mainly due to encroachment of obnoxious exotic aquatic weeds, destruction of wetlands and pollution of water and soil. Numbers of aquatic species are under threat of extinction. These endemic plants have narrow range of distribution, narrow range of tolerance power and are usually very specific in their ecological requirements. Therefore, there is an urgent need to survey, document and collect detailed information on present status, distribution, ecology and taxonomy of these plants which will help in their conservation and utilization for sustainable development.

In present investigation, attempts have been made to survey the aquatic angiosperms of coastal plains Konkan, Goa and Karnataka. In taxonomic treatment citations, full synonymy, botanical descriptions followed by flowering and fruiting months, and taxonomic and ecological notes are given. The herbarium specimens collected during present studies have been cited in the end. The nomenclature, as far as possible, is updated by consulting most recent literature on the aquatic plants.

Review of Literature

Arber (1920) in her book 'Water plants: A study of aquatic angiosperms' survey the older morphological and anatomical literature and the book is indispensable even to the modern students. Sculthorpe (1967) in the 'Biology of aquatic vascular plants' has summarized more recent literature relating to the ecology and biology of aquatic plants. Similarly C.D.K. Cook *et al.* (1974) has done taxonomic overview of fresh water aquatic plants in his book 'Water plants of the world'.

The major work on survey of aquatic flowering plants of India includes Biswas and Calder (1936); they have prepared account of common water and marsh plants of India and Burma. Subramanyanam in 1962 has made major contribution to the knowledge of taxonomy of aquatic plants of India. For first time he provided key for identification of aquatic flowering plants with some excellent illustrations. C.D.K. Cook (1996) in his book 'Aquatic and wetland plants of India has recorded about 485 species of aquatic and semiaquatic flowering plants with keys for identification, descriptions and illustrations.

Several other workers such as Chavan and Sabnis (1961) Satyanarayana (1962), Seerwani (1962), Vyas (1964), Mujumdar (1965), Gupta (1965), Kaul and Zutshi (1967), Manilal and Sivarajan (1978), Sahai and Sinha (1968), Kaushik (1969), Karthikeyan *et al.* (1982), Gopal (1990), Naskar (1990) have contributed to aquatic plant diversity of India. In last few decades additions to the aquatic angiosperms is made by several workers. Number of new species have been added by Bruggen (1968), Raghvan, *et al.* (1982), Joseph and Sivarajan (1988), Joseph (1991), Joseph and Sivarajan (1990), Yadav and Govekar (1994) etc. C.D.K. Cook has revised aquatic flora of India and reported over 485 aquatic and semiaquatic angiosperms. He has also revised genera viz. *Rotala* (1979), *Neehamandra* (1982), *Blyxa* (1983) and *Ottelia* (1984). Sivarajan and Joseph (1983) have revised genus *Nymphoides* for India. In addition to this, several other general publications on floristics and vegetation have made references to aquatic plants.

Although there is considerable amount on general flora of coastal plains of Konkan Goa and Kanara, there is not single source documentation of aquatic plant diversity of the region. Takings into account the importance of wetland and wetland flora the present work was undertaken in year 1996.

Plan of Work and Methods Adopted

The present work has been undertaken mainly to bring-out a comprehensive survey of aquatic plant diversity of coastal plains of Konkan, Goa and Kanara (Map-I and II). The present study also aims to find out some areas of rich biodiversity of aquatic plants in study region for conservation

The present work is an outcome of 6 years extensive and intensive field studies on the study region. During the period several field visits were arranged to the selected localities of area. Number of field trips of 5–6 days duration were arranged to collect plant specimens from different localities. About 50 main localities were selected from study area for collection of plant species. The localities were selected to cover

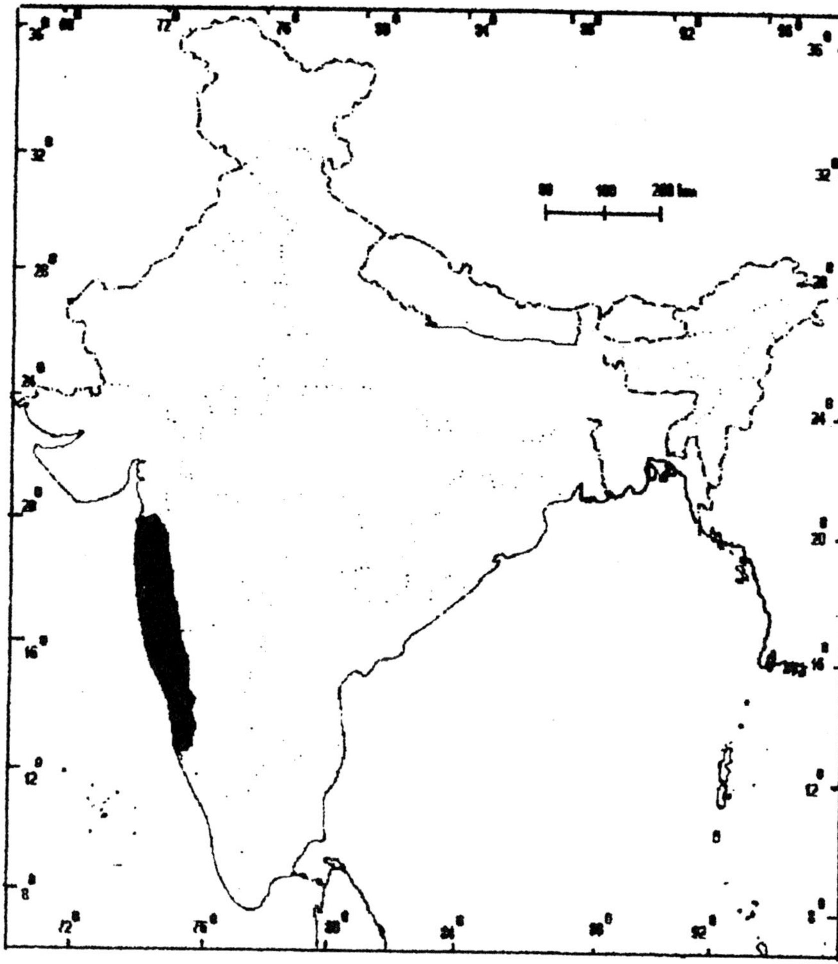

Map I. Showing Study Area, Coastal Plains of Konkan, Goa and Kanara.

the entire study region. During collection special attention was paid to rare, endemic aquatic plants, while collecting the specimens, the data such as habit, habitat, month of flowering and fruiting, ecological note and field note were collected. After returning to the headquarters, all the specimens were processed and poisoned and finally the specimens were processed for herbarium by using usual methodology (Rao and Sharma, 1990). Herbarium specimens are deposited in the Herbarium of Botany Department of Shivaji University.

Most of the specimens and field identifications were confirmed satisfactorily with the help of available literature such as floras,

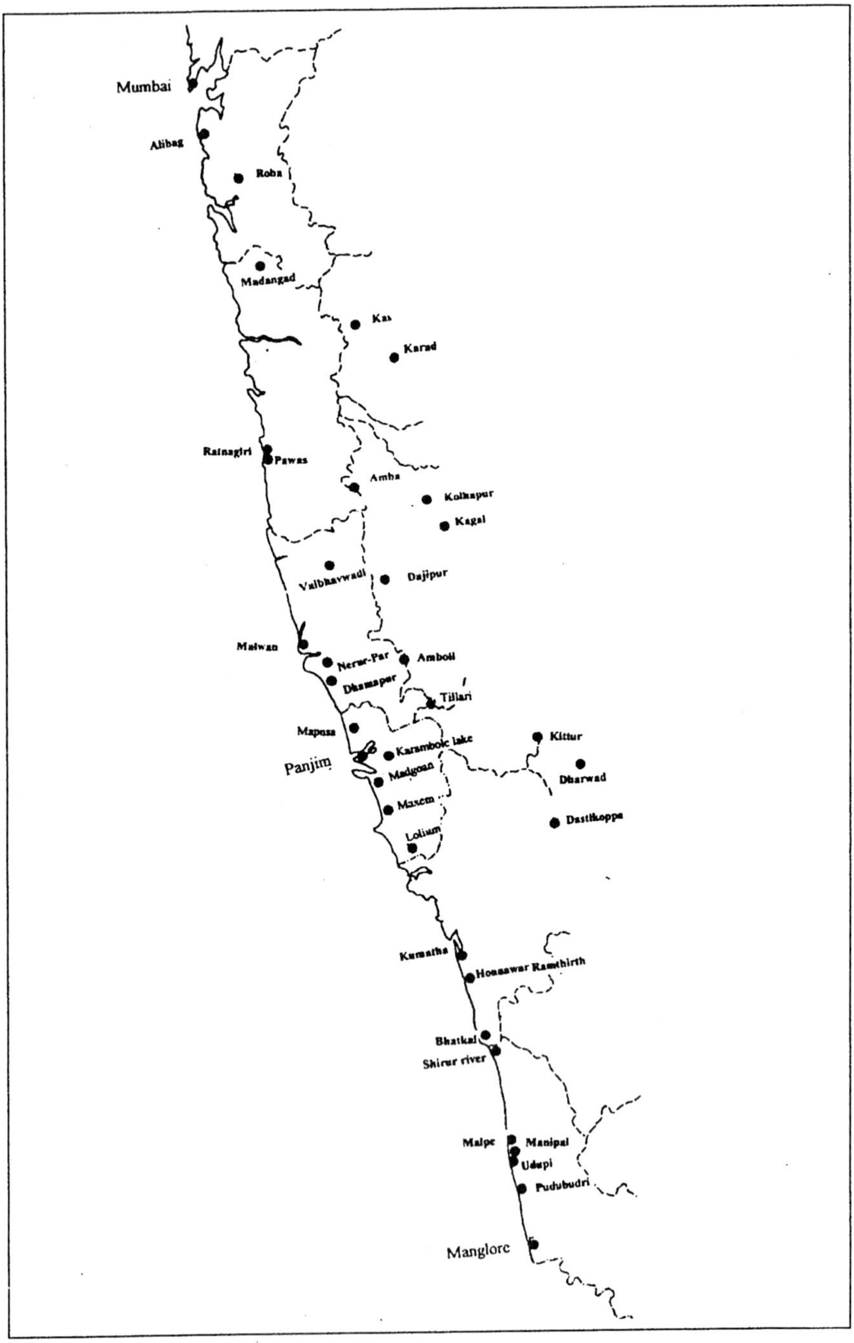

Map II. Showing Localities in the Study Area.

monographs, revisions and research articles. All references have been cited at the appropriate places in the enumeration. The tentatively identified specimens and doubtful identifications were confirmed by matching specimens with authentically identified specimens deposited in the Herbarium of Botanical Survey of India, Western Circle, Pune and Blatter Herbarium, St. Xavier's College, Mumbai. Some specimens that could not be assigned to any Indian species have been referred to expert in the family for their opinion.

The present data based mainly on the information collected during last four years. The families were arranged according to their alphabets in the present work.

The nomenclature adopted in present work is based on latest taxonomic literature and in accordance with the recommendations made by International Code of Botanical Nomenclature (Stafleu, 1972). The correct names of the species followed by a reference to the recent monographs, taxonomic revision and basionym, if any. Each species is provided with citation, description, phenological data, ecological note, occurrence and taxonomic note. Account of the family Eriocaulaceae has already been given in first volume of Book 'Biodiversity of India', therefore, it is excluded from present account of aquatic plants.

Discussion

During intensive and extensive search for aquatic plant diversity for last six years resulted into collection of 244 aquatic and semiaquatic species of flowering plants. There are about 67 true aquatic, 149 semiaquatic, 16 mangroves and 13 ravirine flowering plants found in the coastal plains of Konkan, Goa and Kanara. Of the 244 species 139 belong to dicots and 105 to monocots recorded for the region.

Ponds and puddles on the lateritic plateaus in coastal plains harbour very peculiar and interesting monsoon wetlands. Monsoon wetlands possess very characteristic flowering plants. Coastal plains of Konkan, Goa and Kanara are traversed by number of streams and rivers. These streams are usually temporary and remain dry for most of period in a year. During rainy season, they possess very characteristic aquatic plants of great ornamental values viz. *Eriocaulon setaceum, E. dalzellii, Rotala rotundifolia, Cryptocoryne cognata, C. crispatula* etc.

Riverbeds form unique ecosystem and possess very characteristic vegetation. The important rivarine plants include *Homonoia riparia, Melastoma malabathricum, Phyllanthus lawii, Barringtonia ancutangula,*

Tamarix ericoides, Salix tetrasperma etc. These plant species play important role in stabilization of riverbank and erosion.

General survey of mangrove vegetation revealed that there are about 16 species of flowering plants. The important mangroves includes *Rhizophora mucronata, Sonnertia alba, Kandelia candelabrum, Acanthus ilicifolius, Lummitzera racemosa, Aegiceros corniculata* etc. This is very important ecosystem under great pressure of human activities. There is need for conservation of coastal mangrove ecosystem of the region.

The studies have revealed that the area is fairly rich in aquatic angiosperms. Many of them are of common occurrence but some are of rare viz. *Aponogeton satarensis, A. bruggenii, A. appendiculatus, Eriocaulon ratnagiricus, Indotristicha ramosissima, Nymphoides aurantiacum, Rotala ritchiei* etc.

Acknowledgement

Authors are thankful to Head, Department of Botany, Shivaji University, Kolhapur for providing necessary research facilities and to Department of Science and Technology, Government of India for providing financial assistance.

TAXONOMIC ACCOUNT

True Aquatics

Dicots

ACANTHACEAE

Hygrophila pinnatifida (Dalz.) Sreem. in Bull. Bot. Surv. India 10 (2): 222. 1968; Cook in Aquat. Wetl. Pl. India 33. 1996. *Nomaphila pinnatifida* Dalz. in Hook. J. Bot. 3: 38. 1851. *Cardanthera pinnatifida* Benth. ex Cl. in Hook. f. Fl. Brit. India 4: 405. 1884.

A slender, glandular-pubescent, aromatic herb, 40–60 cm high. Stem 4-gonous, swollen at nodes. Leaves opposite, 5–10 × 2–4 cm, oblong-oblanceolate, glandular-pubescent; lower leaves deeply pinnatifid and oblique. Flower usually solitary in opposite axils; bracts deeply pinnatifid. Calyx divided almost to the base; sepal lobes 5, narrow, linear and usually unequal. Corolla 2-lipped, purple, 2–3 cm long; upper lip with 2-oblong obtuse lobes & lower lip with equally 3 lobed. Stamens 4, fertile with glabrous filaments and anther cells parallel, equal. Ovary narrow, 2-celled with many ovules. Capsules narrow, sessile, pointed:

Fls. & Frts.: Jan.–Apr. *Exsiccata*: 2308

Occasional, rooted submerged hydrophyte.

Occurrence: Dodhamarg, Vaibharwadi. Endemic.

Ecological Note: It grows in running water rocky and gravelly streams, in vegetative stage it is totally submerged in the water.

Hygrophila polysperma (Roxb.) T. Anders. in J. Linn. Soc. 9: 456. 1867; Cl. in Hook. f. Brit. India 4: 406. 1884; Cook in Aquat. Wetl. Pl. India 33. 1996. *Justicia polysperma* Roxb. Fl. Ind. 1: 119. 1832.

A small procumbent herbs, 20–30 cm high; stem much branched with roots at the nodes. Leaves opposite, subsessile, glabrous with tapering base. Inflorescence terminal, spike 3–7 cm long; bracts 6–3 mm, elliptic-oblong with a strong mid-nerve, pubescent; bracteoles linear-lanceolate. Calyx tubular with 5 sepals, pubescent, divided in the middle. Corolla pale to dark blue, 2-lipped, palate of the lower lip, glabrous. Stamens 4, 2 fertile, upper stamens reduced to staminodes. Capsules narrow, oblong, shortly apiculate, glabrous, valves recurved after dehiscence. Seeds 20–30, orbicular on hard upwardly curved retinacula.

Pls. & Frts.: September-February. *Exsiccata*: 2242

Common, rooted submerged hydrophyte.

Occurrence: Malwan, Panjim, Maxem, Karwar, Phonda

Ecological Note: It grows in fast running water rocky and gravelly streams. In vegetative form it is totally submerged in the water, in flowers it emerges from the water.

Hygrophila ringens (L.) Steud. Nom. 1: 418. 1821. *Ruellia ringens* L., Sp. Pl. 635. 1753. *R. quadrivalvis* Buch.-Ham in Trans. Linn. Soc. Lond. 14: 219. 1824; Cook in Aquat. Wetl. Pl. India 34. 1996. *H. quadrivalvis* (Buch.-Ham.) Nees in Wall. Pl. As. Rar. 3: 80. 1832; Cl. in Hook. f. Fl. Brit. India 4: 407. 1884.

Erect, much-branched herb; stem quadrangular, woody at base, hispid. Leaves lower obovate to oblong and upper lanceolate, 5–15 × 1.5–4.5 cm, narrowed at base and tapering into 2–3 cm long petioles, entire, acute or obtuse, glabrous or hispid. Flowers sessile in dense, opposite, axillary, clusters forming distant whorls; bracts oblong, 4–5 mm long, obtuse, hairy; bracteoles lanceolate, 3–3.5 mm long, acute, hairy. Corolla purplish blue, 1.2–1.5 cm long, hairy outside; tube, narrow at base, abruptly widened upwards; limb 2–lipped; lobes oblong. Stamens 4; filaments hairy at base. Capsules terete, brown, 1.2–1.5 cm long, shortly beaked. Seeds 20–24, orbicular, hygroscopically hairy.

Fls. & Frts.: October–February. *Exsiccata*: 2310

Rare, emergent aquatic.

Occurrence: Bhatkal, Kumtha, Nerur-Par.

Ecological Note: It grows on margins of lake.

CERATOPHYLLACEAE

Ceratophyllum demersum L., Sp. Pl. 992. 1753; Hook. f. Fl. Brit. India 5: 639. 1888; Subram. Aq. Angio. 52. 1962; Cook in Aquat. Wetl. Pl. India 84. 1996. *C. verticillatum* Roxb., Fl. Ind. 3: 624. 1832.

Submerged herb, forming a much-branched tangle of slender branches and leaves. Leaves whorled, dichotomously divided into filiform, 2–2.5 cm long, toothed segments. Flowers minute, axillary solitary. Male flowers with 6–12, narrow, subclavate, 2–fid tepals. Stamens 20–30; filaments very short; anthers with terminally produced 2–fid connectives. Female perianth as in males, 1–1.5 mm long. Ovary sessile; style subulate. Fruits ellipsoid, 3.5–4 × 2 mm, excluding spines, slightly compressed, biconvex, with narrow, usually half as long as terminal one.

Fls. & Frts.: Oct.–May. *Exsiccata*: SUK 2210

Common, rooted submerged hydrophyte.

Occurrence: Alibag, Bhatkal Karwar, Manipal Udupi.

Ecological Note: It grows in ponds, lakes throughout the coastal plains of Konkan, Goa and Kanara.

CONVOLVULACEAE

Ipomoea aquatica Forssk., Fl. Aegypt-Arab. 44. 1775; Cl. in Hook. f. Fl. Brit. India 4: 210. 1883; Cook in Aquat. Wetl. Pl. India 92. 1996. NELICHI-BHAJI

Perennial, aquatic herbs; stems hollow, spongy, 2–3 m long, trailing or floating on water, glabrous or hairy at nodes. Leaves variable, often ovate-lanceolate, 8–12 × 4–8 cm, truncate, cordate, sagittate or hastate at base, acute or obtuse and mucronate at apex, glabrous. Flowers solitary or in few-flowered, axillary cymes; peduncles 1.5–8 cm long; pedicels 2–5 cm long; bracts minute. Corolla pink-purple, 3–5 cm long, glabrous. Stamens included; filaments hairy at base. Capsules ovoid, 7–8 mm long, glabrous. Seeds 2–4, broadly 3-gonous, grayish pubescent.

Fls. & Frts.: Sept.–Dec. and Apr.–June. *Exsiccata*: SUK 2315

Abundant, rooted emergent hydrophyte.

Occurrence: Dhamapur, Kolhapur, Pudubudri.

Ecological Note: It grows in ponds, sometimes cultivated for stem and leaves used as vegetable.

HALORAGACEAE

Myriophyllum oliganthum (Wt. & Arn.) F. Van Miller Fragm. Phyt. Austr. 10: 54. 1876; Cook in Aquat. Wetl. Pl. India 214. 1996. *Haloragis oligantha* Wt. & Arn., Prodr. 338. 1834. *Myriophyllum intermedium* Wt., Ill. 2: 23, t. 102. *excl.* f. 8. 1850; Cl. in Hook. f. Fl. Brit. India 2: 433. 1878 *excl. syn.*

Glabrous, aquatic herb. Leaves serrate or pectinately pinnatifid, often whorled. Flowers axillary, monoecious or hermaphrodite. Male flowers: Petals 2–4. Female flowers: petals absent. Fruit with rounded and minutely scabrous on ridges.

Fls. & Frts.: August–February. *Exsiccata*: 2345

Common, emergent aquatic hydrophyte.

Occurrence: Alibag, Panvel, Maxem, Punjim, Udupi.

Ecological Note: It grows in lakes as troublesome weed.

LENTIBULARIACEAE

Utricularia aurea Lour. Fl. Cochinch. 26. 1790; Janarthanam & Henry, Bladd. of India, 30–34. 1992; Cook in Aquat. Wetl. Pl. India, 235–236. 1996. *U. flexuosa* Cl. in Hook. Fl. Brit. India 4: 329. 1884. (Plate-IV)

Herb; rhizoids usually absent, if present always four, verticellate at base of raceme, inflated, fusiform; leaves dissected, primary segments 3–5 per node, semi-verticellate, secondary segments two from a point. Traps ovoid; mouth lateral, oblique with a ring of setae around mouth. Bracts basifixed, one nerved. Pedicels recurved and distally thickened in fruit. Calyx lobes subequal, ovate, fleshy, spread in fruit. Corolla bright yellow; upper lip ovate; lower lip obovate, bigibbous at base. Capsule globose with a long beak; seeds polygonal, margin winged, testa cells more or less isodimetric.

Fls. & Frts.: Oct.–Feb. *Exsiccata*: SUK 554

Frequent, submerged floating herb.

Occurrence: Ratnagiri, Roha, Kumptha.

Ecological Note: Submerged floating in ponds, lakes along with *Hygrorrhiza aristata, Pseudoraphis spinescens, U. stellaris, Limnophila, Blyxa, Nymphoides* species.

Utricularia australis R. Br. Prodr. 430. 1810; Janarthanam & Henry, Bladd. of India, 34–37. 1992; Cook in Aquat. Wetl. Pl. India. 236. 1996.

Submerged floating herb; rhizoids usually absent, if present branched, branches botryform. Leaves dissected, primary segments dichotomo sly divided; ultimate segments setulose. Traps obliquely obovoid; mouth lateral, oblique with two appendages. Bracts basifixed, one nerved. Pedicels erect or spreading in fruit. Corolla bright yellow; upper lip ovate; lower lip transversally elliptic, prominently unigibbous at base. Fruit setting is not seen.

Fls. & Frts.: Oct.–Feb. *Exsiccata*: SUK 563

Rare, floating herbs.

Occurrence: Bhatkal, Khopoli.

Ecological Note: Submerged floating herb in stagnant water of ponds, lakes along with *Hygrorhiza aristata, Pseudoraphis spinescens, Limnophila, Blyxa, Nymphoides* sps. etc.

Utricularia stellaris L. f. Suppl. Pl. 86. 1781; Cl. in Hook. Fl. Brit. India 4: 328. 1884; Janarthanam & Henry, Bladd. of India, 102–106. 1992; Cook in Aquat. Wetl. Pl. India. 244–245. 1996. (Plate-VI)

Aquatic, submerged, floating herb; rhizoids absent. Stolons branched. Leaves dissected; primary segments 3–5, secondary segments 2 per node and bifurcated. Traps obliquely ovoid; mouth lateral oblique. Racemes held above water by floats. Floats globose, ellipsoid, 4–7, verticellate. Bracts basifixed, three to seven nerved. Pedicels erect at anthesis and recurved in fruit. Calyx lobes subequal, fleshy. Corolla yellow, glandular hairy; upper lip ovate; lower lip orbicular, bigibbous at base. Capsule globose, circumcissile; seeds prismatic, 4–7 angled, scarcely winged, testa cells elongate, reticulate at sides and bottom and isodimetric on top.

Fls. & Frts.: Aug.–Dec. *Exsiccata*: SUK 976

Common, submerged aquatic herbs.

Occurrence: Dapoli, Mandangad, Lolium, Maxem, Bincholim.

Ecological Note: Submerged floating in stagnant water of ponds, lakes and rice fields.

LYTHRACEAE

Rotala macrandra Koehen, Bot. Jahrb. 1: 176. 1880 and Engl. Pflanzenfam. IV. 216: 41. 1903; Joseph et Sivarajan in Proc. Ind. Acad. Sci. (Plant Sci.) 99 (3): 185. 1989; Cook in Aquat. & Wetl. Pl. India 258. 1996. *Rotala rotundifolia* sensu Blatt. & Hallb., J. Bombay nat. Hist. Soc. 25: 718. 1918, *p.p. Ammania rotundifolia* Wt. & Arn., Prodr. 1: 306. 1834; Cl. in Hook. f. Fl. Brit. India 2: 566. 1879.

Annual herb; stem creeping and rooting at nodes. Inflorescence on peduncled cylindric racemes, 1–8 together, forming a kind of terminal thyrsus; flowers crowded towards the apex of racemes often lax towards its base. Petals bright rose coloured. Stamens exerted. Capsules 4-valved, slightly longer than broad. Seeds brownish yellow.

Fls. & Frts.: Nov.–Dec. *Exsiccata*: SUK 2234

Common, emergent aquatic herbs.

Occurrence: Kalasepar, Maxem.

Ecological Note: Frequent along margins of streams and lakes.

Rotala ritchiei (Cl.) Koehne, Bot. Jahrb. 1883 and in Engl., Pflanzenr IV. Blatt. and Hallb., J. Bombay Nat. Hist. Soc. 1903. *Ammonia ritchiei* C.B. Cl. in Hook. & Fl. Ind. 2: 5. 1879.

An aquatic, rooted, slender, glabrous, annual, 30–40 cm tall. Stem branched, creeping or floating below, simple obovate. Leaves opposite, decussate, leaves below the water elliptic-lanceolate, rounded at apices while emergent leaves more or less rounded, 3.3 cm long. Bracteoles much smaller than calyx tube. Flower axillary, solitary, much shortly pedicillate; calyx 4-lobed 1.7 mm long, tubular, campanulate, acute; petals four, less than 1 mm long, slightly exceeding the calyx tube, ovate, alternate to the sepals; stamen 4, inserted below the middle of calyx; ovary ellipsoid, style very short, persistent, stigma capitate; Capsule ellipsoid, opening by 4 valves; seeds elliptic, brownish.

Fls. & Frts.: Aug.–Nov. *Exsiccata*: SUK 2237

Endangered, submerged partly emergent herbs.

Occurrence: Kas. Endemic.

Ecological Note: It grows along shallow margin of fresh water streams and ponds.

Rotala rotundifolia (Buch.-Ham. *ex* Roxb.) Koehne. Bot. Jahrb. 1: 175. 1880 and in Engl. Pflanzenf. IV. 216: 41. 1903; Joseph *et* Sivarajan in Proc. Ind. Acad. Sci. (Plant Sci.) 99 (3): 193. 1989; Cook in Aquat. & Wetl. Pl. India 260. 1996. *Ammania rotundifolia* Buch.-Ham. *ex* Roxb. Fl. Ind. ed 1. 1: 446. 1820; Cl. in Hook. f. Fl. Brit. India 2: 566. 1879.

Aquatic or amphibious, heterophyllous, annual herb, branched creeping and rooting below. Leaves all decussate, sessile or shortly petioled, green or reddish. Flowers in terminal simple or branched bracteate spikes, monomorphic bracts closely imbricating, green or pinkish. Petals 4, pink. Stemens 4, nerve exerted the calyx tube.

Fls. & Frts.: Sept.–Mar. *Exsiccata*: SUK 2245

Common, emergent aquatic herbs.

Occurrence: Achirne, Alibag, Bhatkal, Manipal, Udupi.

Ecological Note: It grows in marshes, ditches on plateaus.

MENYANTHACEAE

Nymphoides aurantiaca (Dalz.) O. Ktze. Rev. Gen. Pl. 1: 429. 1891; Subram. in Aquat. Ang. 24. 1962. Sivarajan *et al.*, Aquat. Bot. 145–170. 45. 1993; Cook in Aquat. Wetl. Pl. India 264. 1996. *Limnanthemum aurantiacum* Dalz., J. Bot. Kew Gard. Misc. 2: 136. 1850; Cl. in Hook. f. fl. Brit. India 4: 132. 1893. (Plate-I)

Floating aquatic herb; stem rooting at nodes. Leaves alternate, orbicular, 3–10 cm across, deeply cordate at base, shining green above, purplish beneath. Flowers in dense clusters on nodes, just below the leaves. Calyx divided almost to base, teeth lanceolate, obtuse. Corolla yellow, 1.5–2 cm across, lobes obovate, 6–8 mm long, fimbriate, rounded at apex with a broad longitudinal crest down the middle of each lobe. Stamens 4–7. Ovary unilocular. Capsules ovoid-oblong, 4.5 mm long, beaked. Seeds 5–6, globose, minutely echinulate.

Fls. & Frts.: Sept.–Dec. *Exsiccata*: SUK 2212

Rare, rooted hydrophytes.

Occurrence: Madgoan. Endemic.

Ecological Note: It grows in ricefield.

Nymphoides cristatum (Roxb.) O. Ktze. Rev. Gen. Pl. 429. 1891. *Menyanthes cristata* Roxb., Pl. Corom. 2: 3. t 105. 1798. *Limnanthemum cristatum* (Roxb.) Griseb. Gen. & Sp. Gent. 342. 1839; Cl. in Hook. f. Fl. Brit. India 4: 131. 1883. KUMUDINI (Plate-II)

Free floating aquatic herb; stem rooting at nodes. Leaves alternate, orbicular or orbicular-oblong, 3.5–12.5 cm across, deeply cordate at base, shining green above, purplish beneath. Flowers in dense clusters on nodes, just below the leaves. Calyx divided almost to base, teeth oblong-lanceolate, 5–6 mm long, obtuse. Corolla white, 2–2.5 cm across, lobes obovate, 6–8 mm long, rounded at apex with a broad longitudinal crest down the middle of each lobe. Stamens 4–7. Ovary unilocular. Capsules ovoid–oblong, 4.5 mm long, beaked. Seeds 5–6, globose, minutely echinulate.

Fls. & Frts.: Sept.–May. *Exsiccata*: SUK 2328

Common, rooted hydrophyte.

Occurrence: Alibag, Maxem.

Ecological Note: It grows in water logged ricefield and in stagnant water ponds and puddles.

Nymphoides indicum (L.) O. Ktze. Rev. Gen. Pl. 429. 1891; Cook in Aquat. Wetl. Pl. India 265. 1996. *Menyanthes indica* L., Sp. Pl. 145. 1753. *Limnanthemum indicum* (L.) Griseb. *emend.* Thw. Enum. Pl. Zeyl. 205. 1860; Cl. in Hook. f. Fl. Brit. India 4: 131. 1883. (Plate-II)

Free floating aquatic herb; stem rooting at nodes. Leaves alternate, orbicular or orbicular-oblong, 3.5–12.5 cm across, deeply cordate at base. Flowers in dense clusters on nodes just below the leaves. Calyx divided almost to base. Corolla white, lobes oblong, crested in the middle. Stamens 4–7, filaments short. Ovary unilocular. Capsules globose, 6–8 mm across. Seeds more than 20, bright yellow.

Fls. & Frts.: Oct.–Apr. *Exsiccata*: SUK 2302

Frequent, rooted hydrophyte.

Occurrence: Ratnagiri, Pavas.

Ecological Note: It grows in waterlogged ricefields and lakes.

Nymphoides parvifolium (Griseb) O. Ktze. Rev. Gen. Pl. 2: 429. 1891; Subram., Aqaut. Ang. 24. 1962; Sivarajan *et al.*, Aqaut. Bot. 45: 145–170. 1993; Cook in Aquat. Wetl. Pl. India 266. 1996. *Limnanthemum parvifolium* Griseb. in DC. Prodr. 9: 141. 1845. *Limnanthemum parvifolium* var. *moonii* (Thw.) Cl. in Hook. f. Fl. Brit. India 4: 132. 1883. (Plate–II)

Rooted aquatic, often apparently terrestrial. Rhizome small, erect, with many roots. Branches slender, delicate, upto 20 in long. Leaves of two kinds; one of radical leaves forming rosette at the base of plant other type of leaves terminating the branches. Floating leaves deeply cordate with wide sinus at the base, rounded at the apex. Flowers in cluster. Sepals 4, oblong, acute, 1 mm long. petals 4, oblong, subacute, apiculate, narrow towards base. Seeds 10, orbicular, lentiform, microscopically echinate.

Fls. & Frts.: Sept.–Feb. *Exsiccata*: SUK 2216

Common, rooted hydrophyte.

Occurrence: Bhatkal, Honawar, Panjim.

Ecological Note: It grows in wet ground often along streams.

NELUMBONACEAE

Nelumbo nucifera Gaertn., Fruct. Sem. 1; 73, t. 2. 1788; Subram. Aquat. Ang. 8. 1962; Cook in Aquat. Wetl. Pl. India 272. 1996. *Nymphaea nelumbo* L., Sp. Pl. 511. 1753. *Nelumbium speciosum* Willd., Sp. Pl. 2: 1258. 1799; Hook. f. & Thoms. in Hook. f. Fl. Brit. India 1: 116. 1872. KANKADI

A large herb with rhizomatous root stock. Leaves alternate, orbicular, concave or cupped, erect, peltate, 30–90 cm across, entire, radiately nerved, glaucous; petioles distinctly prickly. Flowers solitary, 12–25 cm across, rosy white or dull yellow. Sepals small. Petals 14–30 cm long, elliptic-oblong, obtuse, spread or suberect. Anthers with a clavate appendage. Torus 2 cm high, spongy; the top flat; 2.5 cm across. Carpels 12–30, 1–1.5 cm long, borne singly in cavities in the pithy obconical receptacles.

Fls. & Frts.: Aug–Sept. *Exsiccata*: SUK 2305

Occasional, rooted hydrophyte.

Occurrence: Alibag, Honnavar

Ecological Note: It grows in permanent water bodies, often cultivated in lakes.

NYMPHAEACEAE

Nymphaea nouchali Burm. f., Fl. Ind. 120. 1768; Cook in Aquat. Wetl. Pl. India 274. 1996. *N. stellata* Willd. Sp. Pl. 2: 1153. 1799; Hook. f. & Thoms. in Hook. f. Fl. India 1: 114. 1872. UPALIA KAMAL

Perennial herb. Rhizomes short, ovoid, acute. Leaves usually floating, entire or bluntly toothed, 10–20 cm in diameter; orbicular-elliptic or submerged sagittate with a 5–7 cm deep sinus, glabrous on both surfaces. Flowers solitary, 6–15 cm in diameter. Sepals lanceolate to ovate-lanceolate. Petals 6-many, pale violet, blue or purple. Stamens

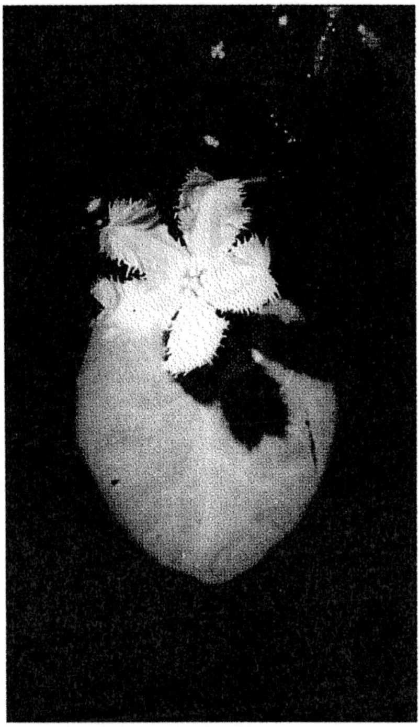

Nymphoides aurantiaca

Nymphoides parvifolium

Nymphoides hydrophylla

Nymphoides indicum

Plate I

Sagittaria guayanensis

Aponogeton hybrid (*A. appendiculatus* × *A. undulatus*)

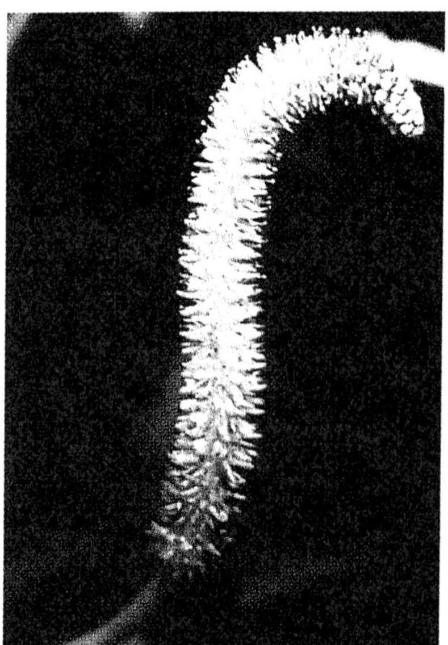

Aponogetonn natans

Aponogeton satarensis

Plate II

Aponogeton bruggenii

Aponogeton bruggenii

Aponogeton undulatus

Aponogeton crispus

Plate III

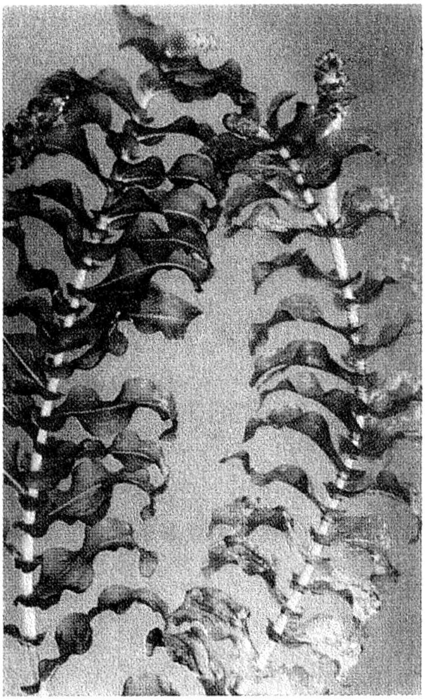

Potamogeton crispus

Potamogeton crispus

Potamogeton nodosus

Potamogeton pectinatus

Plate IV

Utricularia albocerulea

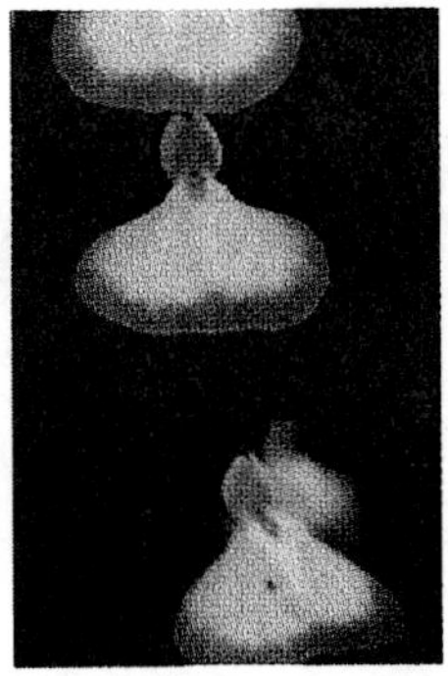

Utricularia purpurascens

Utricularia reticulata

Utricularia stellaris

Utricularia aurea

Utricularia exoleta

Utricularia striatula

Utricularia caerulea

Plate V

Cryptocoryne crisputla

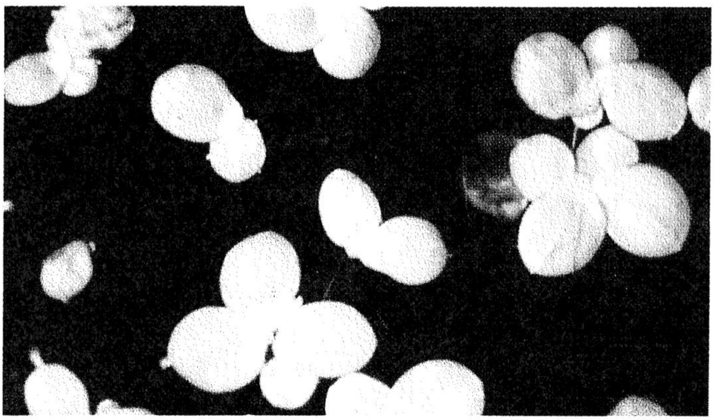

Spirodella polyrrhiza

Trithuria konkanensis

Plate VI

View of lake near Dhampur

View of stream near Pudubudri

Plate VII

many, anthers flat, with a whitish blue aperture at the top. Filaments dilated at the base. Stigmatic disc surrounded by short carpels. Fruit globular. Seeds longitudinally striated.

Fls. & Frts.: Mar.–Dec. *Exsiccata*: SUK 2239

Common, rooted hydrophyte.

Occurrence: Bhatkal, Dapoli, Karwar, Lolium, Malwan, Mangalore, Manipal, Mapusa, Maxem, Panjim, Ratnagiri, Udupi.

Ecological Note: It often dominant in permanent water tanks and lakes. the flower open in morning and close by evening. Once the flower is pollinated, it does not open.

Nymphaea pubescens Willd. Sp. Pl. 2: 1154. 1798; Mitra, Fasc. Fl. India 20: 18. 1990 & in Sharma *et al.* (ed.) Fl. India 1: 431. 1993; Cook in Aquat. Wetl. Pl. India 275. 1996. *N. lotus* auct. non. L. (1753): Hook. f. & Thoms. in Hook. f. Fl. Brit. India 1: 114. 1872, *utreque p.p. N. lotus* var. *pubescens* (Willd.) Hook. f. & Thoms. in Hook. f. Fl. Brit. India 1: 114. 1872. KAMAL

Rhizomes short, ovoid, acute. Leaves peltate, usually floating, entire or bluntly toothed, 10–20 cm in diameter; orbicular, elliptic or submerged sagittate with a 5–7 cm deep sinus, glabrous on upper surface and pubescent on the lower surface. Flowers solitary, 6–15 cm in diameter. Sepals anceolate to ovate-lanceolate. Petals 6 to many, pale violet, blue or purple, sometimes fragrant. Stamens many, anthers flat, with a whitish blue aperture at the top. Filaments dilated at the base. Stigmatic disc surrounded by short carpels. Fruit globular. Seeds longitudinally striated.

Fls. & Frts.: Mar–Dec. *Exsiccata*: SUK 2220

Common, rooted hydrophyte

Occurrence: Karwar, Manipal, Udupi

Ecological Note: It grows in shallow permanent water bodies.

Nymphaea rubra Roxb. ex Andr., Bot. Rep. 8 (104): t. 503. 1808; Mitra, Fasc. Fl. India 20: 21. 1990 & in Sharma *et al.* (ed.) Fl. India 1: 432. 1993; Hook. f. & Thoms. in Hook. f. Fl. Brit. India 1: 114. 1872, *utreque p. p. quoad. syn N. rubra., N. nouchali* auct. non. Burm. f. (1768): Subr., Aqaut. Ang. 6. 1962, *p. p.* SALUK

Perennial herb. Leaves sharply and irregularly sinuate-dentate, velvety pubescent and prominently veined beneath. Flowers solitary,

5–15 cm across. Sepal, oblong obtuse, 5–10-ribbed. Petals 16–25; crimson red, with a dark coloured band in the middle. Stamens 55–80.

Fls. & Frts.: Feb.–Jun. *Exsiccata*: SUK 1622

Rare, rooted hydrophyte

Occurrence: Bhatkal, Kalshe par, Manglore, Udupi.

Ecological Note: Grows in more eutrophicated lakes.

PODOSTEMACEAE

Dalzellia zeylanica (Gardn.) Wt. var. **konkanica** (Wills.) Janardhanan in Bull. Bot. Surv. India 24: 207. 1982; Cook in Aquat. Wetl. Pl. India 317. 1996. *Lawia zeylanica* Tul. var. *konkanica* Wills. in Ann. Roy. Bot. Gard. Perad. 1: 216. 1902.

Annual herb; roots frondose, flat, closely attached to rock, usually irregularly lobed and branched. Shoots very short, simple. Leaves simple, either on upper surface and edges of thalloid root in closely packed rosettes, those in centre filiform to linear, and below ovate or obovate. Flowers solitary or rarely in pairs, pedicellate, subtended by a small cup like structures; spathella absent. Pedicels 2–3 mm long at anthesis becoming upto 1–2 mm long in fruit. Tepals 3, ovate, 1.5–2.5 mm long, united below. Stamens 3 or very rarely 2, free; filaments 2–2.5 mm long. Capsules ellipsoidal, 1–2.5 mm long, 9-ribbed, 3-locular, opening by 3 equal valves. Seeds ovoid, brown.

Fls. & Frts.: Sept.–Feb. *Exsiccata*: SUK 2304

Common, emergent aquatic.

Occurrence: Ranga Fort

Ecological Note: Common in streams along with *Hydrobyopsis sessilis* (Willis) Engles.

Griffithiella hookeriana (Tul.) Warm. Fam. Podost. Afhandl. 6: 13. 1901; Cook in Aquat. Wetl. Pl. India 318. 1996. *Mniopsis hookeriana* Tul. Ann. Sci. Not. Ser. 3: 11. 1015. 1849. *Podostemon hookerianus* Wedd. in DC. Prodr. 17: 74. 1873; Hook. f. Fl. Brit. India 5: 65. 1886.

Annual herb; roots strap shaped or ribbonlike or very rarely all like; attached very closely to rock. Leaves distichous, 3–4 mm long, at the base hooded. Floral buds usually very crowded with 2–6 distichous leaves. Flowers shortly pedicelled; spathella funnel shaped, 2–3 mm long, toothed at apex, splitting irregularly at tip. Pedicels 3–4 mm long at anthesis, 5–6 mm long in fruit. Tepals 2, minute, one on each side of andropodium 1.5–2 long, as long as or longer than ovary, filaments

0.5–0.7 mm long. Stigmas 2, linear, usually simple. Capsules smooth, 2.5 mm long, opening by 2 unequal valves, the large valves persistent. Seeds many.

Fls. & Frts.: Oct.–Dec. *Exsiccata*: SUK 2224

Frequent, emergent aquatic.

Occurrence: Ranga Fort

Ecological Note: Frequent in streams, very polymorphous.

Hydrobryopsis sessilis (Willis) Engler in Engl. & Prantl; Pflanzenfam 18. a: 60. t.53A–C. 1930; Cook in Aquat. Wetl. Pl. India 319. 1996.

Annual herb; roots thalloid, irregular in outline, closely attached to rocks, 0.3–1 cm wide, regularly pinnately branched. Leaves distichous, 1–3 mm long, the lower ones small, flat and apiculate at apex, the upper ones larger. Floral shoots at edge of spathela boat shaped, remaining embedded in the surrounding leaves. Tepals 2, one on each side of anthropodium, linear, as long as or longer than ovary. Stamens 2; anthropodium 1–1.5 mm long; embedded within the surrounding leaves, smooth, obscurely 8-ribbed, 2-locular, opening by 2 unequal valves; stigmas 2, linear. Seeds numerous, yellowish.

Fls. & Frts.: Oct.–Apr. *Exsiccata*: SUK 2259

Frequent, emergent aquatic.

Occurrence: Ranga Fort.

Ecological Note: Common along with **Griffithiella hookeriana** (Tul.) Warming.

Indotristicha ramosissima (Wt.) Van Royen in Acta Bot. Neerl. 8: 474. 1959; Subr. Aqua. Angio., 45. 1962; Cook in Aquat. & Wetl. Pl. Ind. 320. 1996. *Dalzellia ramosissima* Wt. Ic. t. Pl. Ind. Or. 5: 35. 1852. *Terniola ramosissima* Wedd in DC. Prodr. 17: 47. 1873; Hook. f., Fl. Brit. India 5: 63.1886.

Annual herb; roots with disk-like and lobed hold fasts; vegetative shoots arise from the hold fasts upto 45 cm long, branched, bearing scales and ramuli; ramuli and scales, the uppermost 1–3 ramuli and 5–7 leaf like scales from the cup like structure at the base of pedicel; pedicels 7–15 mm long in fruit. perianth segments ovate to obovate, 2–3 mm long in fruit. Perianth segments ovate to obovate, 2–3 mm long, membranous, reddish, united along the lower half. Filaments 3–7 mm long. Capsules ellipsoidal to oblong.

Fls. & Frts.: Nov.–Apr. *Exsiccata*: SUK 2236

Rare, emergent aquatic.

Occurrence: Tillari

Ecological Note: Grows in swift running water.

I olypleurum stylosum (Wt.) J. B. Hall in Kew Bull. 26: 131. 1971; Cook in Aquat. Wetl. Pl. India 323. 1996. *Dicraea stylosa* Wt., l.c. t. 1916f. 2. 1852. *Podostemon stylosus* (Wt.) Benth. in Gen. Pl. 3: 112. 1880; Hook. f. Fl. Brit. India 5: 64. 1886. *P. algaeformis* Benth. in Hook. f. Fl. Brit. India 5: 65. 1886.

Annual herb; roots ribbon like, closely attached to rocks, fleshy, and partly free; irregularly branched. Floral shoots developed at the base of floating parts; at almost regular intervals each becoming 4–5 imbricate and distichous scale like leaves. Flower buds usually as long as broad; pedicels usually 2–5 mm in andropodium. Stigmas 2, linear.

Fls. & Frts.: Sept.–Dec. *Exsiccata*: SUK 2207

Rare, emergent aquatic.

Occurrence: Amboli

Ecological Note: Grow in swiftly running water on rock substratum.

SCROPHULARIACEAE

Limnophila heterophylla (Roxb.) Benth. Scrop. Ind. 25. 1835; Hook. f. Fl. Brit. India 4: 270. 1884; Cook in Aquat. Wetl. Pl. India 346. 1996. *Columnea heterophylla* Roxb. Fl. Ind. (Carey. ed.) 3: 97. 1832.

Annual herb; upper leaves lanceolate, acute 2–8 cm, sessile, 3-nerved, lower leaves pinnatifid, segments narrow, linear. Flowers bluish-white, axillary, solitary, sessile or subsessile. Calyx membranous, 0.5–0.8 cm long; lobes 5, lanceolate, acute. Corolla 0.8–1.2 cm long, 2-lipped; tube cylindric; upper lip suberect, slightly 2–fid, lower lip spreading, 3-lobed Stamens 4, didynamous, anthers disjunct, oblong. Ovary 2-celled, oblong-globose, style slender, deflexed at top, stigma 2-lamellate. Capsules subglobose.

Fls. & Frts.: Aug.–Dec. *Exsiccata*: SUK 2238

Common, emergent aquatic.

Occurrence: Alibag, Bhtakal, Honnavar, Karwar,

Ecological Note: It grows in stagnant waters.

Limnophila indica (L.) Druce, Bot. Exch. Club. Soc. Brit. Isles. 3: 420. 1914; Pennell, Scrop. West. Himal. 26. 1943; Cook in Aquat. Wetl. Pl. India 346. 1996. *Hottonia indica* L., Syst. Nat. ed. 10. 919. 1759. *Limnophila gratioloides*. R. Br. Prodr. 442. 1810. *L. racemosa* Benth. Scrop. Ind. 26. 1835; Hook. f. Fl. Brit. India 4: 271. 1884.

Erect, glabrous herb; often rooting from lower nodes. Leaves whorled, dimorphic, lower submerged, pectinately divided into linear segments, 0.5–2 cm long; segments entire or serrate; the upper leaves entire or 3-lobed, lanceolate, 0.5–2 × 0.3–0.7 cm, serrate-dentate, acute, rarely opposite. Flowers solitary, axillary, passing into terminal racemes; bracteoles linear-subulate, 2–2.5 mm long; pedicels slender, 4–8 mm long, divided half way down; teeth narrowly triangular, acute. Corolla blue, pale purple or white with purple tinge, 7–8 mm long; tube long; limb 2-lipped. Stamens 4; anther cells separate and stalked. Capsules ellipsoid-ovoid, 4–5 × 2–3 mm, glabrous. Seeds many, oblong-quadrangular.

Fls. & Frts.: Aug.–Mar. *Exsiccata*: SUK 2306

Common, emergent aquatic

Occurrence: Achirne, Dapoli, Karwar, Mandangad, Manipal.

Ecological Note: Grow in ponds.

Limnophila polystachya Benth. Scrop. Ind. 76. 1835; Hook. f. Fl. Brit. India 4: 269. 1884; Cook in Aquat. Wetl. Pl. India 348. 1996.

Annual herb; stem erect or ascending which is densely clothed with multifid leaves. Lower leaves capillaceo-multifid; upper leaves opposite or 3–4-nately whorled, linear-oblong or lanceolate, sessile, 2–3.5 × 0.2–1 cm crenulate. Flowers sessile, in terminal and axillary often divaricately branched slender spikes, 5–7.5 cm long; bracteoles 2–3 mm long, linear-lanceolate. Calyx 3–4 mm long, membranous, glabrous, divided about half way down or less, segments deltoid, ovate, acute, subequal. Corolla 0.8 cm long white. Capsules 2–3 mm long, subglobose.

Fls. & Frts.: Dec.–Mar. *Exsiccata*: SUK 2309

Rare, emergent hydrophyte.

Occurrence: Mandangad. Endemic.

Ecological Note: Grow in ponds, paddy fields.

Limnophila sessiliflora (Vahl.) Bl. Bijdr. 849. 1826; Hook. f. Fl. Brit. India 4: 270. 1884; Cook in Aquat. Wetl. Pl. India 350. 1996. *Hottonia sessiliflora* Vahl, Symb. Bot. 2: 36. 1791.

Annual herb; stems glabrous or pubescent. Leaves sessile; the upper often undivided, lanceolate, acute, serrate or dentate, 3-nerved; the lower pinnatisect with numerous very narrow segments. Flowers axillary, solitary, sessile or shortly pedicelled, or terminal arranged in a short spicate racemes. Calyx membranous, rather less than 0.6 cm long, clothed with short gland tipped hairs, divided about 1/2 way down; segments lanceolate, long pointed, very acute. Corolla 0.8–1 cm long, pink. Capsules 0.5 cm long, subglobose, glabrous. Seeds minute, truncate at both ends.

Fls. & Frts.: Oct.–Feb. *Exsiccata*: SUK 2240

Rare, emergent hydrophyte.

Occurrence: Maxem

Ecological Note: It grows in muddy places.

TRAPACEAE

Trapa natans L., Sp. Pl. 120. 1753, var. **bispinosa** (Roxb.) Makino in Iinuma Somoku-Dzusetsu ed. 3, 1: 137. 1907; Cook in Aquat. Wetl. Pl. India 363. 1996. SHINGADA

Stems flexuose. Leaves alternate, crowded on the upper part of the stem, blades more broader than long, 3.5–5 × 3.5–5.5 cm, rhomboid with a triangular apex, somewhat truncate and entire at the base, irregularly inciso-serrate in the upper part, glabrous and shining above and reddish-purple and densely pubescent beneath; petioles terete, dilated in the upper part into a large fusiform spongy float; stipules linear-lanceolate, acute, membranous, caducous. Flowers few, axillary, solitary; pedicels short, stout, hairy; erect in flower and recurved in the fruit. Sepals 4, of which 2 persistent as spines on fruit. Petals 4, pure white, crenulate on the margins. Fruit obovoid, angular, 2–2.5 cm long and broad, with a short conical beak in the centre at the apex and a spreading flattened very sharp spinous horn at either side.

Fls. & Frts.: Dec.–Apr. *Exsiccata*: SUK 2327

Rare in wild, floating hydrophyte. Commonly cultivated.

Occurrence: Alibag, Maxem.

Monocots

ALISMATACEAE

Limnophyton obtusifolium (L.) Miq. Fl. Ind. Bot. 3: 242. 1856; Hook. f. Fl. Brit. India 6: 560. 1893; Subr. Aquat. Ang. 3: 86, t. 3. 1962; Cook, Aquat. Wetl. Pl. Ind. 36. 1996. *Sagittaria obtusifolia* L. Sp. Pl. 993. 1753. NALKUT

Plant with milky latex, upto 75 cm long, robust, scapigerous, stemless. Leaves radical, 15–30 × 10–20 cm, broadly sagittate, pellucid-punctate, acute or subobtuse or with rounded tips; basal lobes long, tapering towards apex to a fine point. Scape 1–3 meter long, stout, angled. Inflorescence with 2–7 whorls of flowers; lowest whorls bearing female flowers. Petals white, obovate, larger than sepals. Stamens 6. Carpels 10–30. Achenes obovoid, base cuneate, with a fine reticulation of scarious ribs, shortly beaked.

Fls. & Frts.: April–September *Exsiccata*: 5306

Rare, rooted emergent aquatic herb.

Occurrence: Manipal, Panjim, Udupi.

Ecological Note: It grows in permanent water bodies.

Sagittaria guayanensis H., B. & K. ssp *lappula* (D. Don) Bogin. in Mem N.Y. Bot. Gard. 9: 192, F. 5. 1955; Subr. Aquat. Ang. 87. 1962; Cook,. Aquat. Wetl. Pl. India 38. 1996. *S. guayanensis* non H. B. & K. 1850; Hook. f. Fl. Brit. India 6: 561. 1893. (Plate-I)

Annual aquatic herbs. Leaves in clusters from the base, floating, petiolate; petiole 30–40 cm long, slender, sheathing at base; lamina 3–10 × 3–7 cm, broadly ovate, many-nerved, deeply cordate at base, rounded at apex. Scape 15–40 cm long, slender, hairy. Flowers large, white, in whorls; lower 1–2 whorls usually with bisexual flowers; upper 1–3 whorls with male flowers, usually each whorl with 3-flowers in lower 1–2 whorls of scapes. Sepals 3, green. thick, chartaceous. Petals 3, white, deciduous. Stamens 6–12, with short filaments; anthers 2-celled, yellow. carpels numerous, free; stigma lateral, papillate. Achene with toothed wings. Male flowers many, 3–5 per whorl in upper portion of scape, pedicel slender, sepals and petals like female flowers; stamens 9–12.

Fls. & Frts.: August–October. *Exsiccata*: 5931

Rare, rooted floating aquatic herb.

Occurrence: Malpe, Mudgaon, Manipal, Udupi.

Ecological Note: It grows in water-logged paddy fields.

Wiesneria triandra (Dalz.) Micheli in Monog. Phan. 3: 82–83. 1881; Hook. f. Fl. Brit. Ind. 6: 652. 1893; Cook, Aqua. Wetl. Pl. Ind. 40. 1996. *Sagittaria triandra* Dalz. in Hook. J. Bot. 2: 144. 1850.

A scapigerous, monoecious herb. Root fibrous. Leaves radical, many, petioled, glabrous; petiole obtusely trigonous, sheathing at the base; lamina narrowly linear spathulate, obtusely keeled at back. Scape erect, simple, terete to obscurely trigonous; flowering axis sharply trignous with unisexual flowers. Lower 2–3 whorls with pistillate flowers, remaining 6–9 whorls with staminate flowers. Female flower shrotly pedicillate; sepals 3, erect, ovate, obtuse; petals 3, altering with sepals. Pistil flask-shaped. Male flower subsessile; sepals 3, ovate-obovate; petals much smaller than sepals; stamens 3. Fruit achene subglobose or obovoid; embryo curved.

Fls.& Frts.: July–September. *Exsiccata*: 5438

Common, rooted emergent aquatic herb.

Occurrence: Lolium plateau, Ratnagiri, Vaibhavwadi and Udupi. Endemic.

Ecological Note: It grows in temporary ponds and puddles on lateritic plateaus.

APONOGETONACEAE

Aponogeton bruggenii Yadav and Govekar, Rheedea 4: 34–36. 1994; Yadav, Aqua Planta 1: 3–12. 1995.; Cook, Aqua. Wetl. Fl. India. 47. 1996. (Plate-III)

Monoecious, tuberiferous, perennial herbs, 10–30 cm in height. Tubers globose to elongate, 2–3 cm in diameter; roots slender, fibrous, from the top of the tuber. Leaves emergent; petiole stout, brittle, 5–25 cm long, acutely triangular; lamina 2–7.5 × 1.8–3.3 cm, thick, dark green, brittle, shiny, ovate-oblong, cordate or rounded at the base, acute or rounded at the apex. Peduncles 20–30 cm long; spike simple, 6–7 cm long, densely flowered. Tepals 2, dark pink, obovate, obtuse, persistent, one-nerved. Stamens 6, exerted; anthers violet, 2-celled. Carpels 3, *ca* 1 × 0.5 mm; stigma decurrent; ovules 2. Follicles 5–7 × 2–3 mm, beaked, smooth; seeds with simple testa; embryo cylindrical, plumule basal, not lying in a groove.

Fls. & Frts.: July–September. *Exsiccata*: 5127

Rare, rooted emergent aquatic herb.

Occurrence: The species was so far known only from type locality Nerurpar-Dhamapur in Sindhudurg district of Maharashtra at about 15–20 m altitude. Recently it has also been collected from Sathose near Savantwadi. Endemic.

Taxonomic Note: Although number of healthy inflorescences were produced in nature, surprisingly, no fruit and seed setting was observed on any inflorescence. Out of the numerous carpels present on the inflorescence, rarely few follicles were seen in few inflorescences.

Aponogeton crispus Thunb. Nov. Gen. 4: 73. 1781 (*'crispum'*); Thwaites, Enum. pl. Zely.: 333. 1864 (*pro nomen*); Hook. f. Fl. Brit. Ind. 6: 564. 1893 (*pro parte*); Bruggen, Blumea 18: 481. f. 1, 6. 1970; Biblio. Bot. 47. 1985; Cook, Aqua. Wetl. Pl. Ind. 48. 1996. *Aponogeton echinatus* Roxb., Fl. India 2: 210. 1832 (*'echinatum'*); Hook. f. Fl. Brit. India. 6:564. 1893; Subr. Aquat. Angio. 92. 1962. (Plate-III)

Monoecious, tuberiferous, perennial, aquatic herbs. Tubers globose-obovoid. Leaves both submerged and floating. Submerged leaves petiolate; petioles bluntly triangular, slender, brittle, sheathing at the base; lamina oblong-lanceolate, elliptic, base rounded or subcordate, apex obtuse. Floating leaves petiolate; petioles slender, terete, brittle, sheathing at the base; lamina very variable in size, ovate-lanceolate, tapering towards apex, apex rounded or blunt, base rounded or subcordate. Spike simple, densely or laxly flowered, white to lilac.Tepals obovate-spathulate, rounded at apex, 1-nerved, white-lilac. Stamens 6, exerted; anthers 2-celled, dark violet, dehiscence longitudinal. Carpels 3; stigma decurrent; ovules 2. Follicles 2-seeded, beaked, smooth or echinate; seeds with simple testa; plumule attached at the base, lying in a groove.

Fls. & Frts.: Throughout year (peak flowering August–November). *Exsiccata*: 5133

Common, Rooted floating hydrophyte with emergent inflorescence.

Occurrence: Ratnagiri, Alibag.

Ecology note: It grows in permanent fresh water ponds, lakes and pools.

Aponogeton natans (L.) Engler & Krause in Engler's, Pflanzenr. 24 (4, 13): 11. 1906; Subra., Aqua. Ang. 92, t. 5. f. 54. 1962; Bruggen, Blumea 18: 477, f. 2. 1970; Biblio. Bot. 50. 1985; Cook, Aqua. Wetl. Pl. Ind. 48 1996; *Aponogeton monostachyon* L. f., Suppl. Pl. 214. 1781. (Plate-II)

Monoecious, tuberiferous, perennial, aquatic herbs. Tubers globose or elongated. Leaves: both submerged and floating. Submerged leaves petiolate, oblong-lanceolate, elliptic, rounded at the base, apex rounded or obtuse, main nerve prominent with 6–8 parallel nerves on either side. Floating leaves petiolate, narrow towards tip, entire, apex blunt or rounded, base cordate, main vein prominent with 5–8 parallel nerves. spike simple, 5–10 cm long, elongated along maturity, densely flowered, lilac-pink. Tepals 2, persistent, oblong-spathulate, 1.5–2 mm long, tip rounded, 1-nerved, pink. Stamens 6, exerted; anther 2-celled, dark blue. Carpels 3. Follicles *ca* 4 × 2 mm, beaked. Seeds 4–8, with double testa; outer testa transparent, inner testa thin, brown, reticulate.

Fls. & Frts.: Almost throughout year (peak flowering period September–November).

Common, Rooted floating hydrophyte with emergent inflorescence.
Exsiccata: 5128

Occurrence: Alibag

Ecology note: It grows in freshwater shallow temporary ponds and permanent lakes.

Aponogeton satarensis Ragh., Kulk. & Yadav, Kew Bull. 36: 687, Fig. I. 1982; Bruggen, Biblio. Bot. 57. 1985; Cook, Aquat. Wetl. Pl. Ind. 50, f. 32. 1996. (Plate-II)

Dioecious, tuberiferous, perennial aquatic herbs, upto 30 cm tall. Tubers 1–2 cm across, globose-ellipsoid. Leaves floating or emergent, petiolate, lanceolate to elliptic, entire, glabrous, shiny, acute or obtuse, main vein prominent with 2–4 parallel nerves on either side. Spike biforked, 4–7 cm long, female spike elongated along maturity upto 13 cm. Male flowers: tepals 2, 1.6–2 × 1.1–1.2 mm, oblong, orbicular to suborbicular, lilac to pink. Stamens 6, in two whorls, exerted; anthers two-celled, dark brown. Pistilode 3, small conical. Female flowers: tepals absent. Carpels 3, 1.8–2 mm long, free; stigma decurrent. Follicles 4–7 × 4 mm, oblong, beaked, pericarp membranous. Seeds 1–2, exalbuminous, elliptic, *ca* 5 × 3 mm, testa simple, plumule lying in a groove.

Fls. & Frts.: June–September. *Exsiccata*: 5129

Endangered, rooted emergent hydrophyte

Occurrence: It is known only from type locality i.e., higher plateaus (900–1200 m) in Satara district. Endemic.

Ecolcgy note: It grows abundantly during June–September as rooted aquatic with floating leaves and emergent biforked spike in temporary ponds and puddles on plateaus.

Taxonomic Note: It is only Asiatic species with dioecious habit and biforked inflorescence.

Aponogeton undulatus Roxb., (Hort. Beng.: 26, 1814, *nomen nud.*) Fl. Ind. ed. Carey 2: 211. 1832 (*'undulatum'*); Bruggen, Biblio. Bot. 41. 1985; Cook, Aqua. Wetl. Pl. Ind. 49. 1996. *Aponogeton microphyllum* Roxb., Fl. Ind. ed. Carey 2: 210. 1832; Hook. f. Fl. Brit. Ind. 6: 564. 1893. (Plate-II)

Monoecious, tuberiferous, perennial, aquatic herbs. Tubers globose-obovoid. Leaves: both submerged and floating. Submerged leaves petiolate, elliptic-lanceolate, narrow towards apex, obtuse or rounded, base rounded sometimes cordate, opaque areas present in irregular pattern between veins, margin undulate. Floating leaves petiolate, oblong, lanceolate, entire, narrow towards apex, obtuse, base rounded or rarely cordate. Propagules (young plantlets) developed on inflorescence like axis. Spathe semipersistent. Spike simple, laxly flowered, white-lilac. Tepals 2, spathulate, clawed, caducous. Stamens 6, exerted. Carpels 3, 0.9–1.2 × 0.4–0.8 mm; ovules 2, basal. Follicles short, terminal, curved beak. Seeds with a simple testa.

Fls. & Frts.: December–May (very seldom in nature). *Exsiccata*: 5130

Common, rooted floating hydrophyte with emergent inflorescence.

Occurrence: Kittur

Ecological Notes: It grows in permanent water bodies such as Lake, Ponds etc.

Hybrid between *A. appendiculatus* and *A. undulatus*. Gaikwad, Sardesai and Yadav, *Aqua Planta* 2: 54–67. 1998. (Plate-II)

Tuberiferous, perennial, robust, submerged aquatic herbs, upto 2 meter tall. Leaves submerged, upto 150 cm long, petiolate; petiole upto 100 cm long, sheathing at the base; lamina lanceolate, *ca* 50 × 6.5 cm, 5–7-nerved, apex rounded, margin undulate, opaque areas present between veins. Spathe *ca* 5 cm long, persistent, acuminate. Inflorescence upto 225 cm long, white–lilac; peduncle upto 180 cm long, cylindrical, thickened towards the spike; spike simple, densely flowered, upto 10 cm long, white–liliac in color, flowers arranged in all directions, old spike greatly elongated at maturity, upto 45 cm long, turning green with age. Flowers: tepals 2, obovate, 2–3 mm long, additional tepals 6-many (upto 20), spathulate, 3-nerved, lilac, persistent. Stamens absent (Probably modified into tepal like structures ?). Carpel 3 to many; ovules two, basal, abortive; no mature follicles and seeds were observed on any inflorescence.

Fls.: September–October. *Exsiccata*: 5324

Rooted submerged hydrophyte with emergent inflorescence.

Occurrence: Pudbudri.

Ecology Note: The stream near Pudbudri in Udupi taluka of South Kanara district.

Taxonomic Note: The hybrid nature of the population is well evidenced by morphological characters. It is like *Aponogeton undulatus* in possessing opaque areas in an irregular pattern between the veins in submerged leaves and formation of plantlets at the tip of inflorescence like axis for the vegetative propagation. It is like *A. appendiculatus* in its habitat preference, geographical distribution, submerged leaves, inflorescence and flower characters. Thus, among all the Indian species of *Aponogeton*, it shares common characters with *A. appendiculatus* and *A. undulatus* (Gaikwad, *et al.*, 1998).

ARACEAE

Cryptocoryne cognata Schott in Bonplandia 5: 222. 1857; Hook. f. Fl. Brit. India 6: 494. 1893; Cook, Aquat. Wetl. Pl. Ind. 55. 1996.

Rhizome vertical, root tuberiferous vermiform. Leaves petiolate; petiole sheathing at the base, as long as the blade; lamina broadly lanceolate, acute, margin undulate. Spadix 9–33 cm long, erect; kettle white; tube short, throat with distinct collar, scarlet red outside, reddish white inside with one or few dark red spots, limb much longer than kettle, twisted, midrib prominent. Male inflorescence separated from female inflorescence by an interspace of 2–5.6 cm. Female inflorescence 0.4–1.3 × 0.25–0.7 cm; 5–6-gynous; ovules 2 seriate, 5–15 per carpel, attached right angle to the central axis; surrounded by many transparent tubular unicellular hairs. Fruit globular, many-seeded. Seeds narrowly elliptical, slightly curved, brownish, black.

Fls. & Frts.: August–November. *Exsiccata*: 5906

Common, rooted submerged or emergent hydrophyte.

Occurrence: Achirne, Malvan, Pavas and Mapusa. Endemic.

Ecological Note: It grows in streams at low altitude.

Cryptocoryne crispatula Engler var. Flaccidifolia N. Jacobsen; Cook, Aquat. and Wett. Pl. Ind. 56. 1996. (Plate-VI)

Rhizome thick, knotty, with stolons, bearing long thick contractile roots and thin absorbing roots. Leaves submerged, linear, ribbon like, upto 50 cm long. spathe limb grey to yellowish with purple to purple black spots, stripes irregular.

Fls. & Frts.: December–January. *Exsiccata*: 5933

Rare, rooted submerged and emergent hydrophyte.

Occurrence: Pudbudri, Shirur river.

Ecological Note: It grows in fresh water rivers and streams.

Lagenandra ovata (L.) Thw. Enum. Pl. Zeyl. 334. 1864; Cook, Aquat. Wetl. Pl. Ind. 56.1996. *Arum ovatum* L. Sp. Pl. 967. 1753. *Lagenandra toxicaria* Dalz. J. Bot. Kew. Gard. 4: 289. 1852; Hook. f. Fl. Brit. India 6: 495. 1893. VATSANAB

Rootstocks reaching *ca* 5 cm thick, creeping, simple; root fibers vermiform. Leaves elliptic-oblong, obtuse or acute, margin undulate, entire, base acute or rounded; petiole as long as the lamina. Male inflorescence cylindric. Female inflorescence of many ovaries. Seeds narrowly oblong, furrowed.

Fls. & Frts.: Feb–May. *Exsiccata*: 5936

Rare, rooted submerged or emergent hydrophyte.

Occurrence: Malvan

Ecological Note: It grows along fresh water streams and ditches.

Pistia stratiotes L. Sp. Pl. 963. 1753; Hook. f. Fl. Ind. 6: 497. 1893; Subr. Aquat. Ang. 75. 1962; Cook, Aquat. Wetl. Pl. Ind. 65. 1996. BRANDAL

Herbs, stemless, stoloniferous; roots of tufted simple white fibers clothed with fibrillae. Leaves obovate-cuneate, rounded or retuse at apex, densely pubescent on both surfaces. Spathes white, obliquely companulate, gibbous, closed below, contracted middle or above, dialated and nearly orbicular above. Fruits membranous, ovoid. Seeds many, oblong or obovoid.

Fls. & Frts.: November–January. *Exsiccata*: 5935

Common, free floating hydrophyte.

Occurrence: Udupi

Ecological Note: It grows in ponds and in permanent water bodies.

CYPERACEAE

Schoenoplectus articulatus (L.) Palla. in Bot. Engl. Jahrb. Syst. 10: 299. 1888; Cook, Aquat. Wetl. Pl. Ind. 175. 1996. *Scirpus articulatus* L. Sp. Pl. 47. 1753; Cl. in Hook. f. Fl. Brit. India 6: 656. 1893. NWALA

Erect, tufted herbs; stem spongy, 1–3 feet high, transversely septate within, flowering near the base than top. Inflorescence capitate. Spikelets ovoid-oblong, rusty-brown, sessile in laterally, stellately spreading clusters. Nuts obovoid, triquetrous, striate with transverse wavy lines.

Fls. & Frts.: June–April. *Exsiccata*: 5322

Common, rooted emergent hydrophyte

Occurrence: Malpe, Panjim, Pawas, Pudbudri.

Ecological Note: It grows in seasonally flooded areas and water logged paddy fields.

HYDROCHARITACEAE

Blyxa auberti Rich. in Mem. Inst. Paris 12, 2: 19, 23, 77, t. 4. 1812; Subr. Aquat. Ang. 59. 1962; Cook, Aquatic Botany 15: 10. 1983; Cook, Aquat. Wetl. Pl. Ind. 216. 1996. *B. oryzetorum* (Decne.) Hook. f. Fl. Brit. India 5: 661. 1888.

Leaves radial, 2.5–60 cm or more long, linear-lanceolate, gradually attenuate a fine point at the apex. Flowers bisexual, sessile, solitary, rarely 1 (2)-flowered, usually emergent. Sepals 6–8 mm long, green, often striped with purple. Petals linear, often twisted, upto 17 mm long, white or reddish. Stamen 3, antisepalous Stigmas linear, 10–15 mm long. Capsule terete. Seeds ellipsoidal, appearing irregularly muricate, apical and basal tails absent.

Fls. & Frts.: July–October. *Exsiccata*: 5327

Rare, rooted submerged hydrophyte.

Occurrence: Panjim, Madgoan, Udupi.

Ecological Note: Grows in temporary water bodies and often in water logged paddy fields.

Blyxa echinosperma (Cl.) Hook. f. Fl. Brit. Ind. 5: 661. 1893; Subr. Aquat. Ang. 59. 1962; Cook, Aquatic Botany 15: 14. 1983; Cook, Aqua. Wetl. Pl. Ind. 217. 1996. *B. lancifolia* Hook. f. *op. cit.*, *B. talbotii* Hook. f. *op. cit.*

An annual herb. Leaves submerged, radical, variable in length, narrowly linear, entire or minutely serrate margin. Flowers bisexual, emergent, longer than the leaves; scape filiform, sepals 3, narrowly linear, obtuse, green. Petals 3, linear, white, often twisted at the top; stamens 3; stigmas 3, laterally papillose. Capsules terete. Seeds spinecent with long lateral tails.

Fls. & Frts.: July–October. *Exsiccata*: 5202

Common, rooted submerged hydrophyte.

Occurrence: Achirne, Pavas, Lolium plateau, Udupi, Pudbudri, Vannawar Ramthirth.

Ecological Note: It Grows in temporary ponds and puddles and water logged paddy fields.

Blyxa octandra (Roxb.) Planch. ex Thw. Enum. Pl. Zeyl. 332. 1864; Subr. Aquat. Ang. 60. 1962; Cook, Aquatic Botany 15: 7. 1983; Cook, Aquat. Wetl. Pl. Ind. 218. 1996. *Vallisneria octandra* Roxb. in Mem. Inst. Paris 12, 2: 77, t. 165. 1802. *Blyxa roxburghii* Rich. in Mem. Inst. Paris 12, 2: 77, t. 5. 1812; Hook. f. Brit. Ind. 6: 660. 1893.

Acaulescent, submerged herbs. Leaves radical, linear-lanceolate, 10–40 × 4.0–0.9 cm, glabrous, membranous, narrow at base, apex attenuate. Flowers dioecious. Male flower: peduncle equal or exceeding the leaves: pedicel 1.2–2.6 cm long; sepals 3, oblong, obtuse; petals 3, oblong; stamen 8–9, filament unequal; anthers subequal; pistilode 3. Female flower: solitary; peduncle 10–12 cm long; ovary equal to spathe. Capsule terete ribbed glabrescent; seeds ovoid-oblong.

Fls & Frts.: July to October. *Exsiccata*: 5321

Rare, rooted submerged hydrophyte.

Occurrence: Lolium.

Ecological Note: Grows in temporary ponds and puddles on lateritic plateaus and in paddy field.

Halophila decipiens Ostenf. in Bot. Tidsskr. 24: 260. 1902; Dixit in J. Bom. Nat. Hist. Soc. 36: 284. 1932; Hartog in Steenis, Fl. Males. 1, 5: 410, f. 18. 1957.

Marine water, dioecious, prostrate herb. Stem thin. Leaves oblong-elliptic, entire, obtuse, base cuneate, covered with short unicellular hairs on one or both sides, 1.0–2.3 × 0.3–0.6 cm. Spathe ovate, with one male and one female flower, tearing at top as fruit ripening. Fruits broadly ellipsoid, beaked, 0.25 × 0.15 cm. Seeds 30, ovate.

Fls. & Frts.: January–February.

Rare, rooted submerged.

Occurrence: Ratnagiri.

Ecological Note: It grows in muddy places in marine water along sea-coast.

Halophila ovalis (R. Br.) Hook. f. Fl. Tasm. 2: 45. 1858; Dixit in J. Bomb. nat. Hist. Soc. 36: 284. 1932; Hartog in Steenis, Fl. Males. 1, 5: 408 f. 16. 1957; Subr. Aquat. Ang. 62. 1962. *Caulina ovalis* R. Br. Prodr. 1: 339. 1810, *Halophila ovata sensu* Hook. f. Fl. Brit. India 5: 663. 1888. *non* Gaud. 1829.

Monoecious creeping, marine water herb. Stem slender. Leaves opposite, linear-oblong to ovate, upto 7.5 × 1.0 cm, palminerved;

stipular scales 2, convolvulate. Spathes sessile, solitary, bracteate. Male flowers pedicelled; female flower usually sessile. Fruit ovate or subglobose, beaked. Seeds ovoid finely reticulate.

Fls. & Frts.: January–February.

Rare, Rooted submerged

Occurrence: Bombay

Ecological Note: It is growing along the seacoast in muddy estuarine region.

Hydrilla verticillata (L. f.) Royle, Ill. Bot. Himal. t. 376. 1839; Hook. f. Fl. India 6: 1893; Cook, Aquat. Wetl. Pl. Ind. 218. 1996. *Serpicula verticillata* L. f. Suppl. 416. 1781. *Hydrilla polysperma* Blatt. in J. Proc. Asit. Soc. Beng. (N. S.) 26: 356. 1931. SHEVAL

Submerged aquatic herb; rooting at nodes. Leaves linear or linear-oblong, apiculate, entire or serrulate. Flowers unisexual, white; male flowers minute, solitary, in subglobose, muricate spathes; female flowers sessile, solitary in cylindrical membranous spathes. Fruit smooth, muricate. Seeds oblong.

Fls. & Frts.: August–February. *Exsiccata*: 5934

Abundant, rooted submerged

Occurrence: Dhamapur, Kolhapur, Panjim.

Ecological Note: It grows in permanent fresh water bodies.

Nechamendra alternifolia (Roxb. ex Wight) Thwaites, Enum. Pl. Zely. 332. 1864; Cook, Aquatic Botany 13: 505. 1982; Cook, Aquat. Wetl. Pl. Ind. 219. 1996. *Vallisneria alternifolia* Roxb. 2: t. 165. 1802 & Fl. Ind. 3: 750. 1864. *Lagarosiphon roxburghii* (Planch.) in Benth. & Hook. Gen. Pl. 3: 451. 1883; Hook. f. Fl. Brit. Ind. 6: 659. 1893.

Submerged herbs. Leaves alternate, becoming clustered at irregular intervals, when clustered they usually distichous, linear. Flowers unisexual. Male flowers arranged in an indeterminate raceme, numerous, densely crowded on inflorescence stalk; sepals 3, free, obovate-orbicular, reflexed; petals obovate, white, reflexed, truncate at apex, irregularly lobed. Stamens 2, fertile, anther bilocular, dehiscing distally. Female flower axillary one or two; sepals 3, free, obovate to orbicular; petals redimentary or absent; ovary unilocular. Fruit narrowly conical to ovoid, indehiscent. Seeds numerous, ovoid but distally elongate.

Fls. & Frts.: August–February. *Exsiccata*: 5927

Occasional, rooted submerged.

Occurrence: Amba, Kittur, Roha.

Ecological Note: It grows in permanent fresh water lakes, pools and canals.

Ottelia alismoides (L) Pers. Syn. Pl. 1: 400. 1805; Hook. f. Fl. Brit. India 5: 662. 1888; Subr. Aquat. Ang. 61. 1962; Cook, Aquat. Wetl. Pl. Ind. 220. 1996. *Stratiotes alismoides* L. Sp. Pl. 535. 1753.

Succulent, flaccid, submerged aquatic herbs. Root fibrous. Leaves usually submerged, very variable, oblong or orbicular, cordate, membranous, margin undulate; petiole 3-gonous. Flower solitary; spathe 5–6-winged, wings undulate, unequal, mouth 5–6-toothed; Peduncles varying in length. Sepals small, oblong, green. Petals nearly orbicular, white with yellowish base, reticulately veined. Ovary narrowly oblong. Fruit ellipsoid, crowded by the withered perianth.

Fls. & Frts.: September–April. *Exsiccata*: 5301

Common, rooted emergent.

Occurrence: Dastikoppa, Kolhapur, Manipal, Panjim.

Ecological Note: It grows in permanent water bodies such as pools, pond, lakes etc.

Vallisneria spiralis L. Sp. Pl. 1015. 1753; Hook. f. Fl. Ind. 6: 660. 1888; Subr. Aquat. Ang. 57. 1962; Cooke, Aquat. Wetl. Pl. Ind. 221. 1996. SAIVALA

Submerged, tufted herbs. Leaves linear, ribbon-shaped, sheathing at base, apex obtuse, margins faintly dentate or entire. Flowers dioecious, on long or short scapes. Male spathes shortly peduncled, ovate. Female spathes connate, tubular. Fruits 5–10 cm long, linear, included in spathes. Seeds numerous, oblong to fusiform, embedded in a gelatinous mass.

Fls. & Frts.: October–April. *Exsiccata*: 5313

Abundant, rooted submerged.

Occurrence: Kolhapur, Dastikoppa, Udupi.

Ecological Note: It grows in permanent fresh water bodies.

LEMNACEAE

Lemna gibba L. Sp. Pl. 970. 1753; Hook. f. Fl. Brit. India 6: 556. 1893; Cook, Aquat. Wetl. Pl. Ind. 227. 1994.

Fronds 0.6–1.2 cm across. Root usually exceeding 3 cm long; root sheath not winged, root cap rounded. Spathe enclosing 2–1 staminate flowers and pistilate. Utricles narrowly winged above, 1–2-seeded. Seeds ribbed and transversely striate.

Fls. & Frts.: September–December. *Exsiccata*: 5938

Common, free floating.

Occurrence: Nerur-Par, Panjim.

Ecological Note: It grows in fresh water ponds, pools and tanks.

Spirodela polyrhiza (L.) Scieid, in Linnaea 13: 392. 1839; Subr. Aquat. Ang. 77. 1962; Cook, Aquat. Wetl. Pl. Ind. 229. 1996. *Lemna maxima* L. Sp. Pl. 970. 1753; Hook. f. Fl. Brit. India 6: 557. 1893. *Spirodela maxima* (Blatt. & Halb.) McC. in J. Bomb. Nat. Hist. Soc. 43: 198. 1942. (Plate-III)

Frond solitary or 2–5 together, reniform, round or obovate with reddish-purple margin. Root many. Stipe hyaline, fugacious. Flowers minute, in marginal clefts of fronds; spathe 2-liped; male flowers in pairs; female solitary. Fruit slightly winged at margin. Seeds 1–2.

Fls. & Frts.: September–April. *Exsiccata*: 5937

Common, free floating.

Occurrence: Nerurpar, Karombal lake.

Ecological Note: It grows in permanent fresh water bodies, forming a thick mat on surface of water.

Wolffia arrhiza (L.) Horkel ex Wimmer, Fl. Schles. ed. 3, 140. 1857; Hook. f. Fl. Brit. Ind. 6: 557. 1893; Subr. Aquat. Ang. 79. 1962; Cook, Aquat. Wetl. Pl. Ind. 230. 1996. *Lemna arrhiza* L. Mant. 294. 1771.

Floating herbs. Frond thick, green, ellipsoidal, 0.4–0.8 × 0.3–0.5 mm. Flowers opening on the surface from simple median pit; stamen solitary, with filament; style short; stigma concave. Utricle ellipsoid, erect.

Fls. & Frts.: June–October. *Exsiccata*: 5939

Occasional, free floating

Occurrence: Kagal

Ecological Note: Grows in permanent fresh water bodies such as lake.

NAJADACEAE

Najas graminea Del. Descr. Egypt. Hist. Nat. 2: 282, t. 50, f. 3. 1813; Hook. Fl. Brit. Ind. 6: 569. 1893; Subr. Aquat. Ang. 100. 1962; Richard,

M. L. Aquatic Botany 24: 176. 1983; Cook, Aquat. Wetl. Pl. Ind. 268. 1996.

Submerged aquatic herbs. Stem slender, smooth, appearing feather like above due to closely packed leaves, rooting at nodes. Leaves linear, obtuse, margin serrulate with spines on each side, 3 cm long; leaf sheath ca. 1 mm long. Male flowers: unisexual, solitary or 2–4 together. Male flowers necked; anther 4-sporangiate. Female flowers naked. Seeds elliptical-oblong.

Fls. & Frts.: June–October. *Exsiccata*: 5312

Common, rooted submerged.

Occurrence: Karambol lake, Kittur, Nerur-par, Udupi.

Ecological Note: Grows in temporary and permanent fresh water bodies such as ponds, pools and lakes.

Najas indica (Willd.) Cham. in Linnea Soc. 4: 501. 1829; Subr. Aquat. Ang. 101. 1962; Cook, Aquat. Wetl. Pl. Ind. 269. 1996. *Najas minor sensu* Hook. f. Fl. Brit. india 6: 569. 1893, *pro majore parte non* All. 1785. *Najas tenuis* R. Br. ex Magnus, Beitr. Najas. 7. 1870 non *Zostera tenuis* Reut. 1854.

Dioecious or monoecious herb. Stem slender, dichotomously branched. Leaves linear, upper recurved, serrulate, auriculate. Flowers fascicled, monoecious, male and female often found in successive leaf axils. Male flowers born in spathe, anthers four sporangiate, elliptical-oblong. Female flower naked. Seeds elliptic-oblong; seed coat pits squarish to hexagonal.

Fls. & Frts.: August-January. *Exsiccata*: 5314

Common, rooted submerged

Occurrence: Karwar, Manipal, Rajapur, Thane.

Ecological Note: It grows in temporary and permanent fresh water ponds, pools and lakes.

Najas marina L. Sp. Pl. 1015. 1753; Cook, Aquat. Wetl. Pl. Ind. 270. 1996; Richard, M. L. Aquatic Botany 24: 162. 1983. *N. major* All. Fl. Pedem. 2: 221. 1785; Hook. f. Fl. Brit. Ind. 6: 569. 1893.

Submerged aquatic herbs, rather robust, brittle. Leaves not auriculate, margin on either side with 5–12 spiny teeth, spines also present on mid-vein. Flower solitary. Male flower born in spathe, 2.5 mm long. Female flower naked. Seeds ellipsoid to ovoid, narrowing to the tip, somewhat compressed. Seed coat pits very irregular in shape.

Fls. & Frts.: November–December. *Exsiccata*: 5940

Occasional, rooted submerged.

Occurrence: Kagal, Kas.

Ecological Note: It grows in permanent fresh water bodies.

Najas minor All. Fl. Predem. 2. 221. 1785. Cook, Aquat. Wetl. Pl. Ind. 271. 1996.

Monoecious, robust, submerged aquatic herb. Leaf truncate to lacerate; auricles serrulate or lacinate with 8–13 spines on each side, none on the inner side, apex rather obtuse. Leaves flat, flexible, serrulate with conspicuous spiny teeth. Male flowers enclosed in spathe, anther 1-sporangiate. Female flower necked. Seeds elliptical-oblong, slightly curved at apex, pitted, seed coat pits rectangular, broader than long.

Fls. & Frts.: June–February. *Exsiccata*: 5932

Common, rooted submerged.

Occurrence: Kittur, Karwar, Karambol lake.

Ecological Note: It grows in permanent fresh water bodies.

POACEAE

Hygroryza aristata (Retz.) Nees ex Wight & Arn. in Edinb. New Phil. J. 15: 380; Hook. f. Fl. India 7: 95. 1896; Cook, Aquat. Wetl. Pl. Ind. 296. 1996. *Pharus aristatus* Retz. Obs. Bot. 5: 23. 1789.

Floating, herbs; culms spongy, rooting at nodes. Leaves linear or ovate-oblong, glaucous beneath. Panicles 3–5 cm long. Spikelets 1–2 cm long; lemmas lanceolate, 5-nerved, hairy on nerves. Grains oblong.

Fls & Frts.: November–March. *Exsiccata*: 5942

Occasional, rooted free floating.

Occurrence: Panjim, Udupi, Dastikoppa.

Ecological Note: It Grows in fresh water lakes and pools

Pseudoraphis spinescens (R. Br.) Vickery in Proc. Roy. Soc. Queensl. 62, n. 7, 69. 1952; Bor, Grass. Ind. 353. 1960; Cook, Aqua. Wetl. Pl. India 311. 1996. *Panicum spinescens* R. Br. Prodr. 193. 1810. *Chamaraphis spinescens* (R. Br.) Poir. in Lam. Encycl. 2: 189. 1811. Hook. f. Fl. Brit. India 7: 62. 1896.

Spikelets 4.5–6 mm long, lacking any subtending bristles of any sort. Lower glumes broadly oblong, 5 mm long, membranous; upper glume acuminate, 4–6 mm long; lemma of upper floret 1.5–3 mm long, membranous, nerveless, glabrous, smooth, somewhat folded around its palea.

Fls. & Frts.: August–September. *Exsiccata*: 5316

Common, rooted emergent

Occurrence: Ratnagiri, Malvan

Ecological Note: It grows in water logged paddy fields.

POTAMOGETONACEAE

Potamogeton crispus L. Sp. Pl. 126. 1753; Hook. f. Fl. Brit. India 6: 566. 1893; Cook, Aquat. Wetl. Pl. Ind. 332. 1996. (Plate-IV)

Submerged aquatic herbs. Stem 30–90 cm long, branched, compressed. Leaves linear-oblong, obtuse, with crisped and finely serrulate margins, translucent; stipule small, caducous. Flowers 2–2.25 mm across, in terminal spike. Tepals suborbicular, clawed, yellowish-brown. Stamen 4; anther sessile. Ovaries 4, one ovulate. Druplet ovoid, compressed, beaked.

Fls. & Frts.: September–February. *Exsiccata*: 5929

Common, rooted submerged.

Occurrence: Kagal, Kittur, Nerur–Par.

Ecological Note: It grows in permanent fresh water bodies such as ponds, pools, lakes and tanks.

Potamogeton nodosus Poir. Encly. Suppl. 4: 535. 1816; Subr. Aquat. Ang. 95. 1962; Cook, Aquat. Wetl. Pl. Ind. 333. 1996. *Potamogeton indicus* Roxb. Fl. Ind. 1: 471. 1820, non Roth ex Roem. et Schult. 1818; Hook. f. Fl. Brit. Ind. 6: 565. 1893. (Plate-IV)

Submerged aquatic herbs. Floating leaves elliptic-lanceolate, oblong or suborbicular, rounded or cuneate at the base, entire, acute, shiny green above, brownish purple beneath; stipule linear, free, membranous. Submerged leaves linear-elliptic, lanceolate, undulate; stipules free. Flowers in leaf apposed dense spike. Tepals suborbicular, clawed. Druplets oblong, beaked.

Fls. & Frts.: September–February. *Exsiccata*: 5315

Common, rooted floating hydrophyte.

Occurrence: Kagal, Kittur.

Ecological Note: It Grows in permanent and temporary fresh water bodies.

Potamogeton pectinatus L. Sp. Pl. 127. 1753; Hook. f. Fl. Brit. Ind. 6: 567. 1893; Cook, Aquat. Wetl. Pl. Ind. 333. 1996. (Plate-IV)

Submerged aquatic herb. Stem filiform, dichotomously branched. Leaves submerged, alternate, narrowly linear, filiform, acute, opaque; Stipules adnate to the leaf sheaths with free tips. Flowers few in distant whorls in an interrupted spike. Peduncle upto 45 cm long. Tepals suborbicular; stamen 4; anther sessile. Ovaries distinct. Druplets turgid, obovoid, slightly compressed, shortly beaked.

Fls. & Frts.: October–February. *Exsiccata*: 5945

Common, rooted floating hydrophyte

Occurrence: Kalamba lake, Nerur-Par.

Ecological Note: It grows in fresh water pools, ponds and tanks.

PONTEDERIACEAE

Eichhornia crassipes (Mart.) Solms in DC Monogr. Phan. 4: 527. 1883; Subr. Aquat. Ang. 70: 1962; Cook, Aquat. Wetl. Pl. Ind. 329. 1996. *Pontederia crassipes* Mart. Nov. Gen. Sp. Pl. 1: 9, t. 4. 1823.

Free floating herb, root fibrous. Leaves obovate, apex rounded or obtuse, base cordate; petiole swollen. Inflorescence spike, enclosed in irregular membranous sheath. Flower lilac or violet, funnel-shaped, trimorphic, heterostyled. Perianth 6, tubular below, lobes unequal, the upper one is larger. Stamen 6, at two levels; filaments with glandular hairs; anthers subequal, dorsifixed. Fruit 3-valved capsule. Seeds numerous.

Fls. & Frts.: July–November. *Exsiccata*: 5947

Abundant, free floating hydrophyte.

Occurrence: Kolhapur, Udupi, Manipal.

Ecological Note: It grows in permanent fresh water bodies such as ponds, pools and lakes.

Monochoria vaginalis (Burm.f.) K.B. Presl, Rel. Haenk. 1: 128. 1827; Hook. f. Fl. Brit. Ind. 6: 363. 1832; Subr. Aquat. Ang. 69. 1962; Cook, Aquat. Wetl. Pl. Ind. 330. 1996. *Pontederia vaginalis* Burm. f. Fl. Ind. 80. 1768. *Monochoria vaginalis* var. *plantagenea* (Roxb.) Solms in DC. Monogr. Phan. 4: 524. 1883; Hook. f. *op. cit.*

Perennial herb. Stem erect or semierect. Leaves broadly ovate, or ovate-oblong, apex acute to acuminate. Racemes 3–25-flowered. Flower dark blue. Capsules 1 cm long, ellipsoid. Seeds brown, oblong, elongate, ribbed, longitudinally, transversely striate.

Fls. & Frts.: May–September. *Exsiccata*: 5946

Common, rooted emergent.

Occurrence: Vaibhavwadi, Pavas, Malpe, Pudubudri, Madgoan.

Ecological Note: It grows in water logged paddy fields and margin of temporary ponds and puddles.

ZANNICHELLIACEAE

Zannichellia palustris L. Sp. Pl. 969. 1753; Cooke, Aquat. Wett. Pl. Ind. 370. 1996. *Z. palustris* ssp. *pedicillata* Hook. f. Fl. Brit. Ind. 6: 558. 1893; Subr. Aquat. Ang. 97. 1962.

Submerged herbs. Leaves alternate or opposite, narrowly linear, filiform, entire. Flowers minute, unisexual. Male flowers necked; stamen 1; connectives prolonged into blunt knob; anther with (2) 4–8 locule. Female flowers born in cup-shaped sheath; style elongate, persisting in fruit. Fruit pithy drupe.

Fls. & Frts.: December–February.

Rare, Rooted submerged.

Semiaquatic Plants

Dicots

ACANTHACEAE

Hygrophila schulli (Buch.-Ham.) M. R. & S. M. Almeida in J. Bombay nat. Hist. Soc. 83 (Suppl.): 221. 1986. *Barleria longifolia* L. in Torner. Cent. II, Pl. 22. 1756. non Nees 1847. *Behel schulli* Buch.-Ham. in Trans. Linn. Soc. Lond. 14: 289. 1825. *Hygrophila spinosa* Anders. in Thw. Enum. Pl. Zeyl. 225. 1860; Cl. in Hook. f. Fl. Brit. India 4: 408. 1884.

Stout herb with numerous fasciculate, usually ubranched subquadrangular stems; nodes tumid, hispid with long hairs. Leaves sessile, oblong-lanceolate, 10–18 × 2–3 cm, in whorls of 6 at each node, the two outer ones much larger than four inner ones. Flowers in axillary clusters of and at each node; bracts lanceolate, hairy and ciliate; bracteoles linear-lanceolate, 1.5–2 cm long, with hyaline margins lower part, hairy and ciliate with long white hairs. Corolla purple-blue, 2.5–3 cm long; tube 1.2–1.5 cm long; limb 2-lipped; lobes oblong, truncate or rounded. Stamens 4, filaments glabrous. Capsules linear-oblong, 7–8 mm long, pointed. Seeds 4–8.

Fls. & Frts.: June–April. *Exsiccata*: SUK 2269

Common, emergent aquatic.

Occurrence: Karwar, Malwan, Phonda.

Ecological Note: It grows in wet and marshy places.

Hygrophila serpyllum (Nees) Anders. in J. Linn. Soc. 9: 456. 1867; Cl. in Hook. f. Fl. Brit. India 4: 406. 1884. *Physichilus serpyllum* Nees in Hook Comp. Bot. Mag. 2: 311. 1837.

A procumbent herb of about 15–30 cm, much branched, stem creeping and rooting at nodes with numerous erect and procumbent branches. Leaves opposite, 0.8–3 × 0.6–1.5 cm elliptic, entire, and more or less hairy. Inflorescence terminal spikes; bracts 0.5 cm long, elliptic, hairy; bracteoles 0.5 cm long, oblong, hairy. Corolla 2-lipped, petals 5, bright blue, palate bright blue with whitish swellings, glabrous. Capsules narrow 0.6 cm long with 10–16 seeds on slender, curved, hard retinacula.

Fls. & Frts.: January–April. *Exsiccata*: SUK 2244

Common, emergent aquatic.

Occurrence: Karwar, Malwan, Panjim.

Ecological Note: It grows in fast running rocky and gravelly streams.

APIACEAE

Centella asiataica (L.), Urb. in Mart. Fl. Bras. 11: 287. 1879; Mukherjee & Const. in Umbel. India 15. 1993. *Hydrocotyle asiatica* L., Sp. Sp. Pl. 234. 1753; Cl. in Hook. f. Fl. Brit. India 2: 669. 1879. BRAHMI

Slender, herbaceous, creeping herb; stems often reddish, rooting at nodes. Leaves orbicular-reniform, rather broader than long. Flowers in fascicled umbels. Each umbel consisting of 3–4, pink, sessile, flowers.

Fls. & Frts.: January–April. *Exsiccata*: SUK 2248

Common, marsh loving.

Occurrence: Karwar, Malwan, Maxem, Panjim.

Ecological Note: It grows in moist places, along gutters, streams and rivers.

Hydrocotyle sibthorpioides Lamk., Encycl. 3: 153; Mukherjee & Const. in Umbel. India 12. 1993: Cook in Aquat. & Wetl. Pl. India 45. 1996. *Hydrocotyle rotundifolia* Roxb. ex DC., Prodr. 4: 64. 1830; Cl. in Hook. f. Fl. Brit. India 2: 668. 1879.

Stoloniferous herb, Petiole 1–7 cm long. Leafblades slightly lobed. Umbel solitary. Petals greenish white.

Fls. & Frts.: May–Dec. *Exsiccata*: SUK 2320

Common, Marsh loving.

Occurrence: Bhatkal, Pawas.

Ecological Note: It grows in moist and shady places and along streams.

ASTERACEAE

Caesulia axillaris Roxb. Pl. Corom. 1: 64. t. 93. 1798; Hook. f. Fl. Brit. India 3: 291. 1881; Pant in Hajra *et al.* (ed.) Fl. India 13: 2. 1995; Cook in Aquat & Wetl. P1. India 67. 1996.

Erect or diffuse, glabrous, subsucculent herb. Leaves sessile linear-lanceolate to oblanceolate, 3–15 × 0.5–2 cm, narrowed at base and then suddenly expanded into membranous, semi-amplexicaulous leaf base, distantly serrulate, acute or subacute. Heads compound, axillary, sessile, containing several uniflowered simple heads, subtended by 2 leafy bracts. Involucral bracts broadly ovate-semiorbicular, winged on the back, produced into scale like appendages at apex. Pappus absent. Achenes obovoid 2–3 mm long, flat, broadly winged due to persistent involucral bracts.

Fls. & Frts.: August–February. *Exsiccata*: SUK 2250

Common, marsh loving.

Occurrence: throughout the region.

Ecological Note: It grows in moist places, rice fields.

Centipeda minima (L.) A. Br. et Ascher., Ind. Sem. P1. Berol. App. 6: 1867. Naithani in Hajra *et al.* (ed.) Fl. 12: 48. 1995; Cook in Aquat & Wetl. Pl. India 68. 1995. *Artemesia minima* L., Sp. P1. 849. 1753. *Centipeda orbicularis* Lour., Fl. Cochinch 493. 1790; Hook. f. Fl. Brit. India 3: 317. 1881.

A small annual herb, stems numerous spreading from the root, prostrate, slender, leafy, usually glabrous. Leaves numerous, alternate, subsessile, 0.6–1 × 0.3–0.5 cm, oblong-spathulate, with few teeth, base tapering. Heads 2–4 mm across, globose, solitary, axillary, subsessile. Involucral bracts small, oblong, obtuse, with membranous margins. Pappus absent. Achenes minute, 4-angled, bristly on the angles.

Fls. & Frts.: Dec–Feb. *Exsiccata*: SUK 2322

Common, marsh loving.

Occurrence: Achirne, Hatkhamba, Pavas.

Ecological Note: It grows in drying ponds, lakes and also in stream & river beds.

Eclipta prostrata (L) L. Mant 2: 286. 1771; Hajra in Hajra *et al.* (ed.) Fl. India 12: 381. 1995. *Verbesina prostrata* L., Sp. Pl. 902. 1753. *Eclipta alba* (L) Hassk. Pl. Jav. Rar. 528. 1848; Hook. f. Fl. Brit. India 3: 304. 1881; Cook in Aquat & Wetland. Pl. India 69. 1996. *Cotula alba* L. Syst. 2: 564. 1767. *Eclipta erecta* L. *op. cit.* MAKA

Erect prostate, much branched annual herb, often rooting at lower nodes. Stems strigose with appressed white hairs. Leaves opposite, sessile or subsessile, oblong-lanceolate or ovate-lanceolate, narrowed to base, entire or faintly serrate, acute, hispid. Heads 5–10 mm across, solitary or 2–7 together, axillary and terminal, on 0.5–5 cm long peduncles. Involucral bracts, 2 seriate ovate-lanceolate, 3–6 mm long, appressedly hairy. Achenes oblong-turbinate or connate, 2 mm long, compressed, with a narrow wing, covered with warty projections.

Fls. & Frts.: Throughout year. *Exsiccata*: 2324

Common, marsh loving.

Occurrence: Kolhapur.

Ecological Note: It grows in wet places.

Epaltes divaricata (L) Cass. Bull. Sci. Soc. Philom. 139. 1818; Hook. f. Fl. Brit. India 3: 274. 1881; S. Kumar in Hajra *et al.* (ed.) Fl. India 13: 147. 1995.

Small, glabrous, annual herb; branches many, spreading, broadly winged. Leaves sessile, alternate, decurrent, linear-oblong, 2.5 × 0.2–0.4 cm, narrowed at base, entire or distantly toothed, obtuse. Heads solitary 5–8 mm across, terminating short winged peduncles. Receptacles convex, naked. Involucral bracts many seriate, lanceolate, rigid, 2–3 mm long, acuminate, the outer shorter, somewhat spreading. Achenes obovoid-oblong, 0.5–1 cm long, black. Pappus absent.

Fls. & Frts.: October–March. *Exsiccata*: SUK 2252

Common, marsh loving

Occurrence: Amba.

Ecological Note: Common on wet places and in harvested rice fields.

Grangea maderaspatana (L) Poir. Encycl. Meth. Suppl. 2: 825. 1811; Hook. f. Fl. Brit. India 3: 247. 1881; Hajra in Hajra *et al.* (ed.) Fl. India 12: 127. 1995; Cook in Aquat. & Wetl. Pl. India 70: 1996. MASHIPATRI

Prostate, annual herb, stems many, spreading, pubescent. Leaves alternate, sessile, pinnatifid, 2–5 cm long, lobes opposite or subopposite, the terminal lobe the largest, all coarsely serrate-dentate, pubescent. Heads globose, 4–8 mm across, 1–2-nate on short, leaf opposed peduncles. Receptacles naked, convex. Achenes subterete, 2–3 mm long, somewhat compressed, glandular. Pappus a short ring of hairs.

Fls. & Frts.: December–May. Exsiccata: SUK 2352

Common, marsh loving.

Occurrence: Karwar, Manglore.

Ecological Note: Common in drying ponds, along margins of lake & river banks.

Sphaeranthus africanus L., Sp. Pl. (ed. 2) 1314. 1763; Hook. f. Fl. Brit. india 3: 275. 1881; S. Kumar in Hajra *et al.* (ed.) Fl. India 13: 159. 1995; Cook in Aquat. & Wetl. Pl. India 71: 1996.

Strongly aromatic, procumbent herb; stem compressed, glabrous, with entire wings; branches numerous, divaricate. Leaves sessile, decurrent, 3.5–6.5 × 1.2 × 2.5 cm, obovate, obtuse, more or less glandular, pubescent finely toothed, narrowed at base. Heads 0.6–0.8 cm across, globose, ebracteate, flowers purple. Involucral bracts obovate-spathulate, rounded at apex, glabrous.

Fls. & Frts.: Oct.–Jan. *Exsiccata*: SUK 2325

Common, marsh loving.

Occurrence: Bhatkal, Honnavar, Kumptha, Manglore.

Ecological Note: It grows abundantly along eastuaries and brackish water.

Sphaeranthus indicus L., Sp. Pl. 927. 1753; Hook. f. Fl. Brit. India. 3: 275. 1881 *p.p.*; S. Kumar in Hajra *et al.* (ed.) Fl. India 13: 160. 1995; Cook in Aquat. & Wetl. Pl. India 72. 1996. GORAKHMUNDI

Strongly aromatic, procumbent herb; branches terete, 15–45 cm long, with toothed wings. Leaves sessile, obovate to lanceolate, 1–6 × 5–2 cm, decurrent to stems, serrate-dentate, obtuse or subacute, glandular-pubescent. Heads compound, spherical, 1–1.5 cm in diameter. Achenes ovoid-oblong, glabrous or pubescent, minute. Pappus absent.

Fls. & Frts.: Nov.–Mar. *Exsiccata*: SUK 2254

Common, marsh loving

Occurrence: Achirne, Ratnagiri, Sawantwadi.

Ecological Note: It grows in drying ditches, harvested rice fields along river and stream banks.

BALSAMINACEAE

Impatiens acaulis Arn., in Hook. Companion Bot. Mag. 1: 325. 1835; Edg. & Hook. f. in Hook. f. Fl. Brit. India 1: 443. 1874; Vivekananthan *et al.* in Hajra *et al.* (ed.) Fl. India 4: 111. 1997.

Perennial herb with tuberous rootstock. Scape 10–20 cm high. Leaves radical, variable in shape and size, 5–20 cm in length and 3–7 cm broad; leaves suborbicular, oblong, obtuse, glabrous, shiny with crenate margin, base rounded and subacute; petioles 3–5 cm long. Flowers pink, 3–7 cm across, on 6–20 cm long racemes; bracts ovate; pedicels ca 7 cm long. Sepals ovate, obtuse; standard short obovate, concave broader upwards. Wings 2-partite with large terminal lobe having a long, slender and curved spur.

Fls. & Frts.: Sept.–Dec. *Exsiccata*: SUK 2256

Common, lithophyte

Occurrence: Amba.

Ecological Note: It grows on steep rock usually out of reach over which water trickles.

BORAGINACEAE

Coldenia procumbens L., Sp. Pl. 125. 1753; Grah. Cat. Pl. Bombay 135. 1839; Cl. in Hook. f. Fl. Brit. India 4: 144. 1883. TRIPANKI

Prostrate or procumbent herb; branches spreading, white, pubescent. Leaves obovate-oblong, 1–2 × 0.5–1.5 cm, tapering at base with crenate and crisped margins, rounded or obtuse at apex, hairy on both surfaces. Flowers solitary, axillary nearly sessile. Corolla white, fading pale yellow. Fruit dry, pyramidal, 4-lobed, 3–4 mm long, separating into one-celled, beaked nutlets; nutlets muriculate hairy.

Fls. & Frts.: August–April. *Exsiccata*: SUK 2356

Common, marsh loving.

Occurrence: Vaibhavwadi

Ecological Note: It grows along margins of lakes and stream.

Heliotropium ovalifolium Forssk. Fl. Aegypt-Arab 38. 1775; Cl. in Hook. f. Fl. Brit. India 4: 150. 1883.

Erect procumbent herb; rootstock woody; stem and branches clothed with long silky and appressed hairs. Leaves elliptic or ovate, 1.5–2.5 × 0.5–0.8 cm, rounded or narrowded at base, entire, obtuse and acuminate, hairy with long silky hairs on both surfaces. Flowers in terminal and leaf opposed dichotomously branched spicate cymes. Corolla white, pubescent outside, 3 mm long, tube cylindric, lobes ovate-oblong. Anthers minute, lanceolate, their connectives produced. Nutlets ovate, 1–1.5 mm long, subacute, densely hairy with silky white hairs.

Fls. & Frts.: August–May. *Exsiccata*: SUK 2330

Common, marsh loving.

Occurrence: Udupi, Manglore.

Ecological Note: It grows along river banks.

Heliotropium supinum L., Sp. Pl. 130. 1753; Cl. in Hook. f. Fl. Brit. India 4: 149. 1883.

Prostrate or procumbent, softly villous herb; stem numerous, spreading, 15–40 cm long. Leaves ovate to ovate-oblong, 1–3 × 0.5–2 cm, acute at base, plicate and crenate along margins, obtuse softly villous on both sides; petioles 3–12 mm long, slender villous. Flowers sessile in simple or forked, 2-sided, densely villous, spicate cymes. Fruit of 2–4 nutlets, 3–4 mm long; nutlets ovoid, sub-compressed, enclosed in large persistent calyx.

Fls. & Frts.: Sept.–May. *Exsiccata*: SUK 2258

Occasional, Marsh loving.

Occurrence: Shiroda

CAMPANULACEAE

Lobelia alsinoides Lam., Encycl. 3: 586. 1791. *Lobelia trigona* Roxb., Fl. Ind. (Carey ed.) 2: 111. 1824; Cl. in Hook. f. Fl. Brit. India 3: 423. 1881.

Small herb; stem prostrate. Leaves subsessile. Flowers purple-blue, axillary, solitary on long pedicels. Stamens 5, basaly ciliate, anthers penicellate. Ovary obconic, 2-celled. capsules ellipsoid, ribbed. Seeds many, brownish, black, trigonous, ribbed.

Fls. & Frts.: Oct.–Dec. *Exsiccata*: SUK 2360

Abundant. marsh loving.

Occurrence: Amboli.

Ecological Note: It grows in drying streams and harvested rice fields.

Lobelia heyneana Schult. in R. & S. Syst. Veg. 5: 50. 1819. *Lobelia trialata* Buch.-Ham. ex D. Don Prodr. Fl. Brit. India 3: 425. 1881.

Small herb; stem winged. Leaves rhomboid-ovate. Flowers bluish-white, axillary, solitary; bracts minute. Calyx lobes 5, triangular, subulate, glabrous. Corolla 2-lobed; upper lobe lanceolate, acuminate; lower lobe spathulate. Stamens 5, anterior 2-anthers penicellate. Ovary hemispheric, 2-celled. Capsules ellipsoid. Seeds ellipsoid, brown.

Fls. & Frts.: Oct.–Feb. *Exsiccata*: SUK 2359

Common, marsh loving.

Occurrence: Throughout the region.

Ecological Note: Frequent in drying streams and harvested rice fileds.

CONVOLVULACEAE

Aniseia martinicensis (Jacq.) Choisy in Conv. Rar. 144. 1834. Convolvulus martinicensis Jacq. Sel. Strip. Amer. 26, t 17. 1763.

Creeping or twinning glabrous herb. Leaves 3.5–7.5 × 1.2–2.5 cm, oblong, entitre, obtuse or emarginate, mucronate, glabrous above, glabrous or more or less pubescent beneath, base acute. Flowers axillary, solitary, with small persistent bracts at the top. Sepals persistent in fruit. Corolla campanulate, 2–3.5 cm long, white, with hairy bands outside. Capsules ovoid, glabrous outside, silky inside. Seeds black, fringed with silky hairs on the margins.

Fls. & Frts.: Aug.–Dec. *Exsiccata*: SUK 2331

Rare, Marsh loving.

Occurrence: Nerur Par.

Ecological Note: Usually creeping in the grasses and in harvested rice fileds.

Ipomoea carnea Jacq. subsp. **fistulosa** (Mart. ex Choisy) Austin in Taxon 26: 237. 1977. *I. fistulosa* Mart. ex Choisy in DC., Prodr. 9: 349. 1845. *I. carnea* Jacq. Enumer. Pl. Carib. 13. 1760.

Erect or scandent shrubs, 2–4 m tall; stems many, fistular, glabrous or pubescent. Leaves ovate to oblong, 12–25 × 5–13 cm, cordate at base, entire, acute or acuminate, minutely pubescent on both surfaces. Corolla pink or entirely white, dark purple within, 8–12 cm across near the mouth. Stamens included; filaments hairy at dilated bases. Capsules ovoid, 1.5–2 cm long, mucronate, pale brown, 4-valved. Seeds 2–4, black, densely brown villous.

Fls. & Frts.: Jan.–Mar. *Exsiccata*: SUK 2257

Common, marsh loving.

Occurrence: Throughout the region.

DROSERACEAE

Drosera L.

Drosera burmannii Vahl, Symb. Bot. 3: 50. 1794; Cl. in Hook. f. Fl. Brit. India 2: 424. 1878.

Perennial herb; stems scanty. Leaves 0.5–1.5 cm long, all radical, rosette, close to ground, spathulate with gland tipped hairs only on upper surface. Hairs on margin long, petioles flattened. Scapes 1–4, erect, glabrous, 4–12 cm high, from the centre of rosette leaves, axillary. Flowers 6–15 on secund racemes. Calyx 4–5 mm, linear-oblong. Petals white, longer than sepals, spathulate. Styles 5. Capsules 5-valved with numerous black, reticulate seeds.

Fls. & Frts.: Novenber–March. *Exsiccata*: SUK 2255

Occasional, marsh loving.

Occurrence: Malwan, Manipal, Pavas.

Ecological Note: Grows in grasslands, paddy fields and near permanent water sources.

Drosera indica L., Sp. Pl. 282. 1753; Cl. in Hook. f. Fl. Brit. India 2: 424. 1878. *D. indica* var. *rubra* Parand. & Diw., Curr. Sic. 11: 304–305. 1943.

Perennial or sometimes annual, stems erect, sometimes branched, glandular, pubescent. Leaves narrow, linear; circinate in vernation, with gland dotted hairs only in the upper surface. Inflorescence leaf opposed racemes, ca 5 cm long, 5–15 flowered; sepals 5 mm long, oblong, lanceolate, subacute. Petals white to rosy coloured, oblong-oblanceolate, longer than sepals. Ovary 3-carpellary with 3 styles. Capsules 5-valved with minute, strongly ridged,

Fls. & Frts.: October-February. *Exsiccata*: SUK 1777

Common, marsh loving.

Occurrence: Malwan, Manipal, Pavas.

Ecological Note: It grows in some water logged places, wet grassland and near permanent water sources.

ELATINACEAE

Bergia ammannioides Roxb. [Hort. Beng. 34. 1814. nom.nud.] ex Roth, Nov. Pl. Sp. 219. 1821; Roxb. Fl. Ind. 1832, *ammaniodes;* T. Dyer in Hook. f. Fl. Brit. India 1: 251. 1874; Bhattacharya in Sharma *et al.* (ed.) Fl. India 3: 33. 1993; Cook in Aquat. & Wetl. Pl. India 184. 1996.

Annual erect, much branched herb; stems reddish purple, terete glandular-pubescent, swollen at nodes. Leaves 15–25 × 3–7 mm, oblanceolate or obovtae-oblong, alternate at base, acute at apex; upper sparsely serrate with gland tipped, teeth lower half, hispid to glandular. Petals reddish-pink, ovate, elliptic or oblong, obtuse. Stamens usually 5, anthers minute. Ovary subglobose, sulcate, glabrous; stigmas thick capitate. Capsules subglobose, reddish. seeds numerous, minute, reticulated.

Fls. & Frts.: August-April. *Exsiccata*: SUK 2263

Common, marsh loving.

Occurrence: Ratnagiri

Ecological Note: It grows on the margins of tanks and in harvested rice fields.

Bergia capensis L., Mant. Pl. 2: 241. 1771; Bhattacharya in Sharma *et al.* (ed.) Fl. India 3: 34. 1993; Cook in Aqaut. & Wetl. Pl. India 185. 1996. *B verticellata* Willd., Sp. Pl. ed. 4, 2: 770. 1790; T. Dyer in Hook. f. Fl. Brit. India 1: 252. 1874.

Glabrous herb, with succulent stems, rooting at nodes; stems terete with pink or reddish striations and constricted nodes. Leaves 2.5 × 0.5–1.5 cm, narrowly elliptic or lanceolate, oblong to oblanceolate, decurrent at base, acute at apex, glabrous. Flowers many in dense axillary cymes, 2.5 mm across. Capsules 2–2.5 mm across, subglobose with 5 longitudinal furrows, septicidal. Seeds numerous, oblong, shining and strongly reticulate.

Fls. & Frts.: August–October. *Exsiccata*: SUK 2262

Common, marsh loving.

Occurrence: Throughout the region.

Ecological Note: Occasional on the margins of tanks, in paddy fields.

FABACEAE

Aeschynomene americana L., Sp. Pl. 713. 1753; Rudd., Reinwardtia 5: 25. 1959; Dasanayake in Fl. Ceylon 7: 165–66. 1991.

An erect or trailing and floating, perennial, shrub. Stem sponge, glabrous, branches few. Stipule glabrous, 1.5 cm. Leaves sessile or shortly pedicellate, imparipinnate; rachis ca 12 cm long, narrow towards tip; leaflets in ca 35 pairs, oblique at the base, linear, oblong, obtuse, apiculate, 1-nerved, glabrous. Inflorescence axillary, 3–4-flowered, corymbose, shorter than the leaves. Pod ca 8 cm long, woody, often indented, acute, straight; born on 1.5 cm long stalk.

Fls. & Frts.: Aug.–Dec. *Exsiccata*: SUK 1815

Common, marsh loving

Occurrence: Panjim, Sawantwadi

Ecological Note: It grows on margins of lakes.

Aeschynomene aspera L., Sp. Pl. 713. 1753; Baker in Hook. f. Fl. Brit. India 2: 152. 1876.

An erect or trailing and floating, perennial, shrub. Stem sponge, glabrous, branches few. Stipule glabrous, 1.5 cm. Leaves sessile or shortly pedicellate, imparipinnate; rachis ca 12 cm long, narrow towards tip; leaflets in ca 35 pairs, oblique at the base, linear, oblong, obtuse, apiculate, 1-nerved, glabrous. Inflorescence axillary, 3–4 flowered, corymbose, shorter than the leaves. Bract ovate, acute, mucronate, ciliate. Pod ca 8 cm long, woody, often indented, acute, straight; born on 1.5 cm long stalk.

Fls. & Frts.: Jul.–Dec. *Exsiccata*: SUK 1834

Category: Marsh loving.

Status: Common.

Occurrence: Alibag, Kalse Par, Nerur Par, Roha.

Ecological Note: Grows on margins of lakes.

Aeschynomene indica L., Sp. Pl. 713. 1753; Baker in Hook. f. Fl. Brit. India 2: 151. 1876.

An erect or trailing and floating, perennial, shrub. Stem sponge, glabrous, branches few. Stipule glabrous, 1.5 cm. Leaves sessile or shortly pedicellate, imparipinnate; rachis ca 12 cm long, narrow towards tip; leaflets ca 35 pairs, oblique at the base, linear-oblong, obtuse, apiculate. Inflorescence axillary, 3–4-flowered, corymbose, shorter than leaves; peduncle hairy, ca 3 cm. Pod ca 8 cm long, woody, often indented, acute, straight; born on 1.5 cm long stalk. Seeds kidney shaped.

Fls. & Frts.: August–December. *Exsiccata*: SUK 1837

Common, marsh loving.

Occurrence: Dapoli, Khopoli, Mandangad, Raigad.

Ecological Note: It grows in ditches and paddy fields.

Geissaspis cristata Wt. & Arn, Prodr. 218. 1834; Baker in Hook. f. Fl. Brit. India 2: 141. 1876; Cook in Aquat. & Wetl. Pl. India 208. 1996.

A procumbent, annual trailing herb; branches terete, glabrous. Leaves abruptly pinnate, petioles 0.4–0.6 cm long; stipules scarious, lanceolate, spured at the base, strongly nerved. Leaflets 4, obovate, 0.5–1.5 × 0.3–0.6 cm, slightly oblique, obtuse, glabrous on both surfaces, narrowded at base; petiolules very short. Flowers in heads; peduncles axillary; pedicels short. Slender curved. Pods 1–2-jointed about 0.4–0.6 am long.

Fls. & Frts.: September–December. *Exsiccata*: SUK 2264

Commn, marsh loving.

Occurrence: Vaibhavwadi.

Ecological Note: Common in grasslands, ricefields, wet places.

GENTIANACEAE

Hoppea dichotoma Willd. in Gen. Naturnf. Fr. Neue. Schr. 3: 434. 1801; Cl. in Hook. f. Fl. Brit. India 4: 100. 1883; Cook in Aquat. Wetl. Pl. Ind. 1996.

Glabrous dichotomously branched glabrous herb. Stems acutely 4-angled. Leaves sessile, ovate or obovate, 4–7 × 1.5–4 mm acute. Flowers in dense terminal dichotomous cymes pedicels short, bracts lanceolate, 1–2 mm long. Corolla pale yellow, tubular, 2 mm long, lobes ovate-oblong, half as long as tube, obtuse.

Fls. & Frts.: October–March. *Exsiccata*: SUK 2287

Common, marsh loving.

Occurrence: Achirne, Kumtha, Honnavar.

Ecological Note: It grows in wet places amidst grasses.

Hoppea fastigata Cl. in Hook. f. Fl. Brit. India 4: 100. 1883; Cook in Auquat. Wetl. Pl. India 1996.

Dichotomously branched glabrous herb; stems acutely 4-angled. Leaves sessile, ovate or obovate, 4–7 × 1.5–4 mm, acute. Flowers in dense terminal dichotomous cymes, pedicels short, bracts lanceolate 1–2 mm very long. Corolla pale yellow, tubular, 2 mm long, lobes ovate oblong, half as long as tube, obtuse. Capsule globose, seeds minute.

Fls. & Frts.: October–March. *Exsiccata*: SUK 2268

Common, marsh loving.

Occurrence: Manglore, Udupi.

Ecological Note: It grows in wet places amidst grasses.

LOGANIACEAE

Mitrasacme indica Wt., Ic. t. 1601. 1850. *M. pusilla* Dalz., Kew J. Bot. 2: 136. 1850. *M. alsinoides* R. Br. Prodr. 543. 1810; Cl. in Hook. f. Fl. Brit. India 4: 80. 1883.

Annual herb, stem slender. Leaves sessile, sheathing and connate at the base, oblong or linear, glabrous. Flowers white, solitary, or 2–3 together clustered in the upper axils. Capsules globose, surrounded by persistent calyx and with 2 curved horns at the apex.

Fls. & Frts.: August–September. *Exsiccata*: SUK 2285

Occasional, marsh loving.

Occurrence: Manipal, Manglore, Satara.

Ecological Note: It grows in wet places amidst grasses.

HYDROPHYLLACEAE

Hydrolea zeylancia (L.) Vahl. Symb. Bot. 2: 46. 1791; Cl. in Hook. f. Fl. Brit. India 4: 133. 1883. *Nama ceylancia* L., Sp. Pl. 226. 1753.

Erect or decumbent, annual herb often rooting at the nodes. Leaves 2.5–6.5 × 0.3–1.5 cm, lanceolate or linear lanceolate, acute. Flowers numerous in racemes on short lateral branches; pedicels 0.3–0.8 cm long, glandular hairy, bracts leafy. Capsule 0.4–1–2 cm, ovoid-oblong, enclosed in enlarged persistent sepals. Seeds numerous.

Fls. & Frts.: November–May *Exsiccata*: SUK 2348

Occasional, marsh loving.

Occurrence: Ratnagiri, Vaibhavwadi.

Ecological Note: It grows in wet places and in harvested rice fields.

LAMIACEAE

Pogostemon deccanensis (Panigr.) Pers. in Bull. Brit. Mus. Nat. Hist. 10: 73. 1982. *Eusteralis deccanensis* Panigr. in Phytologia 32: 475. 1978. *P. tomentosa* Dalz. in Hook. J. Bot. 2: 337. 1850; Hook. f. Fl. Brit. India 4: 641. 1885.

A low herb; stems numerous, erect, simple or branched. Leaves many in whorl, sessile, 0.4–1 × 0.75–1 cm, linear, subacute, more or less pubescent, entire with revolute margins. Flowers sessile in dense pubesent or tomentose spikes, 2.5–6.5 × 0.4 cm. Nutlets ellipsoid, smooth, yellowish-brown.

Fls. & Frts.: September–February. *Exsiccata*: SUK 2271

Common, marsh loving.

Occurrence: Ratnagiri, Vaibhavwadi.

Ecological Note: It grows on plateaus, rice fields and ditches.

LENTIBULARIACEAE

Utricularia albocaerulea Dalz. in Hooker's J. Bot. Kew Gard. Misc. 3:279. 1851; Cl. in Hook. Fl. Brit. India 4: 330. 1884; Janarthanam & Henry, Bladd. of India, 27–29. 1992; Cook in Aquat. Wetl. Pl. India, 233–234. 1996. (Plate-V)

Herbs. Foliar organs ca 1 × 0.4 cm, spathulate; 3-nerved. Traps 1–1.5 mm in diameter, dimorphic, stalk 0.3 mm, appendaged, glandular. Racemes erect, 8–15 cm long, simple, 1–3 flowered, scales basifixed, lanceolate to ovate, truncate at the base, acute at the apex. Flowers 7–11 mm long; pedicels 4–8 mm long, recurved in fruit. Corolla dark blue, yellow in the throat; upper lip 4–4.5 × 3–3.7 mm, obovate, truncate at the apex; lower lip 7–9 × 9–12 mm, ovate hairy along throat, white, bigibbous at base, truncate at the apex; spur 6–9 mm long, slightly curved, conical. Capsules 4 × 3 mm, compressed, ovate or orbicular in outline.

Fls. & Frts.: September–October. *Exsiccata*: SUK 2253

Occasional, monsoon aquatic.

Occurrence: Karwar, Lolium, Malwan, Panjim, Ratnagiri. Endemic.

Ecological Note: It grows on lateritic plateaus.

Utricularia caerulea L., Sp. Pl. 18. 1753; Janarthanam & Henry, Bladd. of India, 42–46. 1992; Cook in Aquat. Wetl. Pl. India. 237–238. 1996.

Herbs. Foliar organs ca 4–7 × 1–1.2 cm, spathulate, 1–6 per node; 1-nerved. Traps 1–1.5 mm in diameter, ovoid with a rim of glandular hairs. Racemes simple, erect. Flowers upto 5 mm long; pedicels upto 2 mm long, erect, recurved in fruit. Corolla variously coloured; upper lip 2–4 × 1.5 mm, p oblong, ciliate at the lower margins; lower lip 1.5–4 × 2–4 mm, broadly ovate, yellow in throat. Spur 3–5 mm long, curved upwards, horizontally projected. Capsules 1.5 × 2 mm, compressed, obliquely ovoid.

Fls. & Frts.: August–December. *Exsiccata*: SUK 2233

Common, monsoon aquatic.

Occurrence: Lolium, Malwan, Ratnagiri, Vaibhavwadi.

Ecological Note: It grows on lateritic plateaus.

Utricularia graminifolia Vahl, Enum. Pl. 1: 195. 1804; Janarthanam & Henry, Bladd. of India, 57–61. 1992; Cook in Aquat. Wetl. Pl. India. 239–240. 1996.

Herbs. Foliar organ linear, 3-nerved. Traps subglobose, subdimorphic; on leaves sessile and elsewhere stalked; mouth basal with two simple, subulate, glandular appendages. Racemes erect; scales and bracts basifixed, one or three nerved. Pedicels slightly winged, erect at anthesis as well as in fruit. Corolla blue streaked, crested at center; lower lip obovate, hairy in throat, bigibbous at base; spur conical, curved, acute at apex. Capsule ovoid, dorsiventrally compressed, thickened along dehisced margin.

Fls. & Frts.: August–June. *Exsiccata*: SUK 722

Common, marsh loving.

Occurrence: Kolhapur, Vaibhavwadi

Ecological Note: It grows margin of perennial streams, in marshy places, soil accumulated on rocks along with grasses.

Utricularia janarthanamii Yadav, S.R., Sardesai, M.M. and Gaikwad, S.P., *Rheedea* 10(2): 107–109. 2000.

Herbs. Foliar organ spathulate, rounded at apex, scattered along stolon. Flowers in racemes on peduncle and solitary on leaves and stolons. Flowers bracteate; bracts basifixed, 1-nerved; pedicel erect at anthesis and recurved in fruit. Corolla blue to pink. Capsules obliquely ovoid, wall uniformly membranous. Seeds numerous.

Fls. & Frts.: July–December. *Exsiccata*: 233

Rare, Marsh loving

Occurrence: Kagal, Kolhapur. Endemic.

Ecological Note: It grows in wet places among grasses.

Utricularia malabarica M.K. Janarthanam & A.N. Henry in J. Bom. Nat Hist. Soc. 86 (1): 84. 1989; Janarthanam & Henry in Bladd. of India, 68–69. 1992.

Herbs. Leaves 3-nerved, obovate. Traps globose; mouth basal with two simple subulate, glandular appendages. Racemes 2–5 cm long, 1–4 flowered; bracts basifixed; pedicels winged, erect in anthesis and recurved in fruit. Corolla blue with white tinge; upper lip truncate at apex; lower lip suborbicualr, bigibbous at base and hairy in throat; spur slender acute at apex. Capsule ovoid, uniformly membranous.

Fls. & Frts.: August–December. *Exsiccata*: SUK 289.

Occasional, monsoon aquatic.

Occurrence: Pavas, Roha, Ratnagiri, Vaibhavwadi. Endemic.

Ecological Note: It is a Common species growing in accumulated soil on lateritic plateaus of Konkan along with grasses.

Utricularia minutissima Vahl Enum. Pl. 1: 204. 1804; Janarthanam & Henry in Bladd. of India, 72–74. 1992. Cook in Aquat. Wetl. Pl. India, 241. 1996.

Herbs. Foliar organ linear, 1-nerved. Traps subglobose; mouth lateral, circular with one subulate appendage. Racemes 2–3 cm long, 1–4 flowered. Corolla pink with white tinge; upper lip oblong, ciliate along margins, rounded; lower lip shallowly 3-lobed, raised at base and hairy in throat; spur straight, horizontal, notched at apex. Capsule obliquely ovoid, compressed, uniformly membranous.

Fls. & Frts.: August–September. *Exsiccata*: SUK 294

Common, monsoon aquatic.

Occurrence: Achirne, Bhatkal, Dapoli, Manipal, Pavas, Ratnagiri. Endemic.

Ecological Note: It grows on lateritic plateaus where wet sandy soil accumulated.

Utricularia naikii Yadav, S.R., Sardesai, M.M. and Gaikwad, S.P., *Rheedea* 10(2): 110–111. 2000.

Herbs. Foliar organs linear-spathulate, 3-nerved. Traps subglobose; mouth basal with two simple, subulate, glandular appendages; glands on appendages sessile and elsewhere stalked. Racemes erect. Corolla blue to pink; upper lip oblong; lower lip obovate, hairy in throat, bigibbous. Capsule ovoid, walls uniformly membranous.

Fls. & Frts.: July–September. *Exsiccata*: SUK 1911

Rare, marsh loving.

Occurrence: Amboli. Endemic.

Ecological Note: It grows amidst grasses on slightly denudated soil.

Utricularia praeterita Taylor in Matthew, Fl. Tamilnadu Carnatic 3: 1120. t. 87. 1983; Janarthanam & Henry, Bladd. of India, 81–83. 1992; Cook in Aquat. Wetl. Pl. India. 242. 1996

Herb. Foliar organ spathulate-linear, 3-nerved. Traps subglobose; mouth basal with two simple subulate glandular appendages. Racemes erect, slightly grooved; scales and bracts basifixed, one-nerved. Corolla violet; upper lip oblong constricted near middle; lower lip obovate, hairy in throat, unigibbous at base; spur descending, acute at apex. Capsule ovoid, dorsiventrally compressed, uniformly membranous; seeds ellipsoid; testa reticulate, elongated.

Fls. & Frts.: July–January. *Exsiccata*: SUK 763

Common, monsoon perennial

Occurrence: Achirne, Shiroda, Vaibhavwadi. Endemic.

Ecological Note: It grows on lateritic plateaus where gravely soil is accumulated

Utricularia purpurascens Graham., Cat. Pl. Bombay 165. 1839; Janarthanam & Henry in J. Econ. & Tax. Bot. 10:245. 1987 et Bladd. of India, 85–89. 1992

Herbs. Foliar organs linear-spathulate, 3-nerved. Traps subglobose; mouth basal with two simple subulate glandular appendages. Racemes erect or very rarely twinning. Corolla pink, white at base; yellow in throat; upper lip oblong; lower lip orbicular, hairy in throat, unigibbous at base; spur sickle shaped curved below lower lip. Capsule ovoid, dehisce marginally thickened vertical slit.

Fls. & Frts.: August–January. *Exsiccata*: SUK 538

Occasional, marsh loving.

Occurrence: Manipal, Ratnagiri. Endemic.

Ecological Note: It grow on plateaus of higher altitude in accumulated soil on rocky substratum.

Utricularia reticulata Smith. Exot. Bot. t. 1805; Cl. in Hook. Fl. Brit. India 4: 331. 1884; Janarthanam & Henry, Bladd. of India, 91–95. 1992; Cook in Aquat. Wetl. Pl. India, 242–243. 1996. (Plate-V)

Herbs. Foliar organs linear, one-nerved. Traps subglobose; mouth basal with two simple subulate appendages. Racemes twinning; bracts basifixed, one-nerved. Pedicels winged, erect at anthesis and recurved in fruit. Corolla violet to pink; upper lip obovate, broader than calyx lobe; lower lip orbicular, hairy in throat, unigibbous at base; spur conical, slightly curved, acute at apex. Capsule ovoid, dehisce by thickened walls; seeds ellipsoid; testa reticulate, elongated, finely striated within.

Fls. & Frts.: August–September. *Exsiccata*: SUK 978

Common, marsh loving.

Occurrence: Bhatkal, Manipal, Ratnagiri, Vaibhavwadi.

Ecological Note: It grows in various habitat.

Utricularia striatula Smith. in Rees, Cyclop. 37: n. 17. 1819; Janarthanam & Henry, Bladd. of India, 106–112. 1992; Cook in Aquat. Wetl. Pl. India. 245. 1996. (Plate-V)

Herbs. Foliar organ orbicular, rosulate at base of scape and scattered. Traps globose; mouth lateral with hairy, glandular appendages. Racemes erect, zigzag; bracts medifixed. Corolla pink to white; often yellow in throat; upper lip deltoid, bidentate; lower lip 3 or 5 lobed; spur conical. Capsule obliquely ovoid.

Fls. & Frts.: July–February. *Exsiccata*: SUK 233

Common, lithophytic and terrestrial marsh loving.

Occurrence: Amba.

Ecological Note: It grows on tree trunks and on dripping rocks.

Utricularia uliginosa Vahl, Enum. Pl. 1: 203. 1804; Sant. in J. Bombay nat. Hist Soc. 49: 218.1950; Janarthanam & Henry, Bladd. of India, 113–116. 1992; Cook in Aquat. Wetl. Pl. India. 246–247. 1996.

Herbs. Foliar organs linear, 3-nerved. Traps subglobose; mouth basal with two simple, subulate, glandular appendages. Racemes erect often grooved and winged. Corolla blue to pink; upper lip oblong; lower lip obovate, hairy in throat, unigibbous. Capsule ovoid, walls uniformly membranous.

Fls. & Frts.: July–September. *Exsiccata*: SUK 237

Occasional, monsoon aquatic.

Occurrence: Manipal, Manglore, Udupi.

Ecological Note: It grows on lateritic plateaus among grasses

LYTHRACEAE

Ammania baccifera L., Sp. Pl. 120. 1753; Cl. in Hook. f. Fl. Brit. India 2: 569. 1879; Cook in Aquat. & Wetl. Pl. India 249. 1996. *Ammania salicifolia* Hiern. in Olive. Fl. Trop. Afr. 2: 278. 1871 *excl. syn.* Monti, 1767; Cl. *loc. cit.* BHARJAMBHAL

Erect, stout herb; branches terete or obtusely 4-gonous. Leaves sessile oblong lanceolate, 3–6 × 0.3–1.2 cm gradually smaller upwards, obtuse or subacute. Flowers numerous densely clustered in axils. Capsules globose red irregularly circumcises.

Fls. & Frts.: August–November. *Exsiccata*: SUK 2272

Common, marsh loving.

Occurrence: Kolhapur.

Ecological Note: It grows in muddy ponds, puddles and pools.

Ammania multiflora Roxb., Fl. Ind. 1: 447. 1820; Cl. in Hook. f. Fl. Brit. India 2: 570. 1879; Cook in Aquat. Wetl. Pl. India 250. 1996.

Herbs. Leaves sessile, opposite, linear-lanceolate to oblanceolate or elliptical, subauriculate, leaves those on stem are larger than those on branch. Solitary, axillary or in 3-many flowered cyme, peduncles upto 10 mm long. Corolla petals 4, obovate, coducous, violet. Stamens 4. Capsule globose, partly exerted or remaining enclosed in a calyx tube, usually red.

Fls. & Frts.: July–December. *Exsiccata*: SUK 2364

Common, marsh loving.

Occurrence: Pawas, Vaibhavwadi.

Ecological Note: It grows in temporary ponds and puddles on lateritc plateaus.

Rotala densiflora (Roth ex Roem. et Sch.) Koeh. in Engl. Bot. Jahrb. 1: 64. 1880 et in Pfreich. 4 (216): 35. 1903; Joseph et Sivarajan in Proc. Ind. Acad. Sci. (Plant Sci.) 99 (3): 181. 1989; Cook in Aquat. & Wetl. Pl. India 255. 1996. *Ammania densiflora* Roth *ex* Roem. et Schult. Syst 3: 304. 1818. *A. pentandra* Roxb., Fl. Ind. 1: 425. 1820; Cl. in Hook. f. Fl. Brit. India 2: 568. 1879.

Erect, annual herb; branches 4-gonous. Leaves decussate. Flowers sessile, soliatry in axils of bracts. Petals deep pink or red. Stamens 5. Capsules globose or ellipsoid, 3-valved. Seeds semi-ellipsoid, light green.

Fls. & Frts.: August–January. *Exsiccata*: SUK 2336

Common, masrh loving.

Occurrence: Achirne, Bhatkal, Mandangad, Roha.

Ecological Note: It grows in marshes and ditches on plateaus.

Rotala fimbriata Wt., Ic. Pl. Ind. Or. 1. t. 217. 1840; Joseph et Sivarajan in Proc. Ind. Acad. Sci. (Plant Sci.) 99 (3): 182. 1989; Cook in Aquat. & Wetl. Pl. India 255. 1996. *Ammania pentandra* Roxb. var. *fimbriata* (Wt.) Cl. in Hook. f. Fl. Brit. India 2: 569. 1879, *excl. syn. A. hexandra* Wall. *A. heyneana* Wall. Cat. 2104, 1828, nom. nud.

Annual herb; stems erect, sometimes creeping, weakly 4-angled. Leaves decussate. Floral bracts auriculate at base. Flowers monomorphic, sessile. Seeds semi-ellipsoid.

Fls. & Frts.: August-November. *Exsiccata*: SUK 2303

Common, marsh loving.

Occurrence: Malvan, Ratnagiri.

Ecological Note: It grows in marshes and ditches on lateritic plateaus.

Rotala floribunda (Wt.) Koehne in Verh. Bot. Vereins prov. Brandeburg, 19 (sitz. 41) 49, 1877, Bot. Jahrb. 1: 156, 1880 and in Engl. Pflanzenf. IV 216: 32. 1903; Joseph et Sivarajan in Proc. Ind. Acad. Sci. (Plant Sci.) 99 (3): 183. 1989; Cook in Aquat. & Wetl. Pl. India 255. 1996. *Nimmonia floribunda* Wt., Madras J. Lit. Sci. 5: 312, t. 20. 1837. *Ameletia floribunda* (Wt.) Wt. in Hook., Ic. Pl. Ser. 2, 5. t. 826. 1840. *Ammania floribunda* (Wt.) Cl. in Hook. f. Fl. Brit. India 2: 567. 1879.

Annual herbs. Leaves alternate, opposite or in clusters, submerged leaves linear; aerial leaves linear to lanceolate. Flowers in dense, terminal, globose racemes, dimorphic, heterostylic, subsessile or pedicelled. Petals 4, showy bright rose. Stamens 4, inserted on lower third of sepal tube. Capsules opening by 2 valves.

Fls. & Frts.: August-December. *Exsiccata*: SUK 2274

Rare, marsh loving.

Occurrence: Mandangad. Endemic.

Ecological Note: It grows along margins of tanks.

Rotala indica (Willd.) Koehne, Bot. Jahrb. 1: 172, 1880 and Engl. Pflanzenf. IV. 216: 40. 1903; Joseph *et* Sivarajan in Proc. Ind. Acad. Sci. (Plant Sci.) 99 (3): 184. 1989; Cook in Aquat. Wetl. Pl. India 257. 1996. *Ameletia indica* (Willd.) DC., Mem. Soc. Phys. Nat. Hist. Geneve 3: 82, t. 3A, 1826 and Prodr. 3: 76. 1828. *Ammania nana* Roxb., Fl. Ind. 9ed. 1) 1: 448. 1820, non Wall. *Ameletia polystachya* Wall. *ex* Wt. & Arn. Prodr. 1: 304. 1834. *A. peploides* Spreng. Syst. 1: 444. 1825; Cl. in Hook. f. Fl. Brit. India 2: 566. 1879.

Erect ascending, glabrous herb, creeping and rooting at base. Leaves decussate, cuneate or narrowed at base. Bracts dimorphic, leaf-like on major branches, which ultimately look like bracteate spikes. Flowers sessile axillary, solitary. Calyx pink or red, appendages absent. Petals 4, pink elliptic. Capsules ellipsoidal, 2-valved.

Fls. & Frts.: October–January. *Exsiccata*: SUK 2366

Rare, marsh loving.

Occurrence: Lolium, Mapusa,

Ecological Note: It grows in marshes, along streams and in rice fields.

Rotala malampuzhensis R. Vasudevan Nair (J. Bombay nat. Hist. Soc. 72: 57. 1975 nom invalid) *ex* Cook in Boissiera 29: 98. 1979; Joseph et Sivarajan in Proc. Ind. Acad. Sci. (Plant Sci.) 99 (3): 186. 1989; Cook in Aquat. & Wetl. Pl. India 258. 1996.

Amphibious or terrestrial annuals. Stems branched, creeping and rooting below. Bracts leafy. Flowers monomorphic, sessile solitary. Petals 3, small bright red. Stamens 3, inserted above the base of calyx tube; nectar scales 3, prominent and alternating with stamens. Seeds 10–15, smooth, bright red.

Fls. & Frts.: August–November. *Exsiccata*: SUK 2337

Common, marsh loving.

Occurrence: Pawas, Vaibhavwadi.

Ecological Note: It grows in marshes and ditches on plateaus.

Rotala occultiflora Koehne, Bot. Jahrb. 1: 152, 1880 and in Engl. Pflanzenf. IV 216: 30. 1903; Joseph et Sivarajan in Proc. Ind. Acad. Sci. (Plant Sci.) 99 (3): 191. 1989; Cook in Aquat. & Wetl. Pl. India 259. 1996.

Amphibious, annual herb. Stems creeping and rooting below. Leaves in whorls of 3. Bracteoles 2, bract-like, longer than flowers. flowers monomorphic, solitary in axils of bractiform leaves. Calyx tube translucent, appendages absent. Petals absent. Stamens 2. Capsule ellipsoidal, 3-valved. Seeds 10–15, semi-ellipsoidal.

Fls. & Frts.: Aug.–Nov. *Exsiccata*: SUK.2339

Common, Marsh loving

Occurrence: Pawas, Ratnagiri

Ecological Note: It grows in moist and wet places.

Rotala rosea (Poir.) Cook in Boissiera 29: 86. 1979; Joseph et Sivarajan in Proc. Ind. Acad. Sci. (Plant Sci.) 99 (3): 192. 1989; Cook in Aquat. & Wetl. Pl. India 260. 1996. *Ammania rosea* Poir. in Lamk, Encycl. Meth. Bot. Sp. 1: 329. 1810. *A. pentandra* Roxb. Fl. Ind. ed. 1, 1: 448. 1820; Cl. in Hook. f. Fl. Brit. India 2: 568. 1879.

Amphibious or terrestrial, annual herb. Stems erect, simple or branched. Leaves decussate. Flowers monomorphic, sessile and solitary in axils of bractiform leaves. Petals 5, pink, obovate, dentate at apex. Stamens 5, inserted just below the middle of calyx tube, included. Capsules globose, 3-valved, exceeding the calyx. Seeds semiovoid, brownish.

Fls. & Frts.: August–December. *Exsiccata*: SUK 2340

Occasional, marsh loving.

Occurrence: Bhatkal, Karwar.

Ecological Note: It grows in ditches along road.

Rotala serpyllifolia (Roth) Bremekamp, Acta. Bot. Neerl. 3: 149. 1954; Joseph et Sivarajan in Proc. Ind. Acad. Sci. (Plant Sci.) 99 (3): 194. 1989; Cook in Aquat. & Wetl. Pl. India 261. 1996. *Micranthus serpyllifolius* Roth, Nov. Pl. Sp. Halberstadt. 282. 1821. *Ameletia tenuis* Wt. Icon. t. Pl. Ind. Or. 1: 13, t. 2578. 1840. *Ammania tenuis* (Wt.) Cl. in Hook. f. Fl. Brit. India 2: 567. 1879.

Amphibious herb. Stems branched from the base, creeping and rooting below. Leaves decussate in distant pairs, submerged leaves linear, aerial leaves sessile or subsessile. Petals longer than calyx lobes, pink. Capsule 2-valved ellipsoid, included in persistent calyx.

Fls. & Frts.: October–February. *Exsiccata*: SUK 2372

Common, marsh loving.

Occurrence: Honnavar.

Ecological Note: It grows on wet rocks in river beds, streams and lakes.

ONAGRACEAE

Ludwigia hyssopifolia (Don.) Exell. Garcia de Orta 5: 471. 1957. *Jussiaea hyssopifolia* G. Don. Gen. Syst. 2: 692. 1832.

Small herb. Leaves alternate, lanceolate, base tapering, apex acuminate. Flowers axillary, solitary. Capsules cylindric. Seeds ellipsoid.

Fls. & Frts.: September–December. *Exsiccata*: SUK 2343

Frequent, marsh loving.

Occurrence: Mapusa.

Ecological Note: It grows in muddy places.

Ludwigia octovalvis (Jacq.) Raven subsp. **sesssiliflora** (Mic.) Raven in Kew Bull. 15: 476. 1962 & Reinwardtia 6: 362. t. 3. 1963. Oenothera octavalvis Jacq. Enum. Syst. Pl. 19. 1760. *Jussiea octanervia* var. *sessiliflora* Micheli in Mant. Fl. Barz. 13: 180. 1875. *J. suffruticosa* L., Sp. Pl. 388. 1753. *Ludwigia suffruticosa* Wall. 1788; Cl. in Hook. f. Fl. Brit. India 2: 587. 1879.

Robust herb. Leaves subsessile, lanceolate, base and apex tapering. Flowers axillary, solitary; calyx tube subquadrangular; lobes four, ovate-elliptic, hairy; petals broadly obovate. Capsules ribbed.

Fls. & Frts.: August–October. *Exsiccata*: SUK 2278

Common, marsh loving.

Occurrence: Kolhapur

Ecological Note: It grows in moist places.

POLYGONACEAE

Polygonum barbatum L. subsp. **barbatum** var. **stagninum** (Buch.-Ham. ex Meissn.) Steward. Contr. Gray. Herb.5: 54. 1930. *P. stagninum* Buch.-Ham. ex Meissn. in Wall. Pl. As. Rar. 3: 56. 1832; Hook. f. Fl. Brit. India 5: 37. 1886.

Erect or ascending, annual or perennial herbs. Leaves sub-sessile, oblong-lanceolate, 5–20 × 1.5–3 cm, acute at both ends, entire, appressed pubescent; ochrea 2–2.5 cm long, with long ciliate mouth. Flowers in terminal, 1.5–2 cm long spikes. Bracts 2.5–3 mm long, ciliate. Perianth white or tinged with pink, 2.5–3 mm long; segments oblong-rounded. Stamens 6–8, shorter than perianth. Nuts trigonous 2–2.5 mm long, smooth, black.

Fls. & Frts.: September–January. *Exsiccata*: SUK 2299

Common, marsh loving.

Occurrence: Kolhapur

Ecological Note: It grows in marshy places and along gutters.

Polygonum glabrum Willd. Sp. Pl. 2: 447. 1799; Hook. f, Fl. Brit. India 5: 34. 1886. *Persicaria glabra* (Willd.) Gomez de la Maza in Ann. Inst. Segunda Ensef. Habana 2: 278. 1896.

Procumbent, perennial herbs. Leaves lanceolate, 7–24 × 1–3 cm, tapering at base, entire, finely acuminate, glabrous, gland dotted; petioles 3–12 mm long; ochreae 2–3 cm long, conspicuously veined, truncate and ciliate at mouth. Flowers in paniculate, slender racemes; peduncles glabrous. Perianth pink, 3–5 mm long; segments oblong, obtuse. Nutlets broadly ovoid or suborbicular, compress biconvex, black, shining.

Fls. & Frts.: September–April. *Exsiccata*: SUK 2282

Abundant, Marsh loving.

Occurrence: Kolhapur.

Ecological Note: It grows near water sources.

Polygonum hydropiper L. var. **glandulissima** Gauge. in Rec. Bot. Surv. India 2: 401. 1903.

A glabrous, annual herbs, 30–45 cm high. Leaves subsessile or shortly petiolate, lanceolate or oblong-lanceolate, very variable in width; stipules glabrous, not ciliate. Flowers in filiform interrupted decurved racemes; bracts glabrous, glandular. Perianth pink, very glandular. Nutlets trigonous, opaque, granulate.

Fls. & Frts.: September–February. *Exsiccata*: SUK 2293

Infrequent, marsh loving.

Occurrence: Kudal, Nerur-par, Sawantwadi.

Ecological Note: It grows near water sources.

Polygonum plebejum R. Br. Prodr. 420. 1812; Hook. f, Fl. Brit. India 5: 27. 1886 (*incl. vars.*).

A diffusely branched, prostrate, marsh loving herb; often with woody rootstock. Leaves oblong or linear, sessile; stipule hyaline, short, lacerate to the middle, fimbriate. Flowers pink, axillary, 2–3 together; pedicels short; perianth broad or narrow, outer two acute. Nut 3-gonous, smooth, shiny

Fls. & Frts.: August–May. *Exsiccata*: SUK 2297

Abundant, marsh loving.

Occurrence: Kagal, Kolhapur.

Ecological Note: It grows on marshy land, in drying ponds and lakes.

RUBIACEAE

Dentella repens (L.) Forst. Char. Gen. P1. Ins. Mar. Austr. 25, t. 13, 1776; Hook. f. Fl. Brit. India 3: 42. 1880; (Repr. ed.); Subram. & Sharma, Bull. Bot. Surv. India 10: 386. 1968; Cook in Aquat. & Wetl. P1. India 335. 1996.

A small, slender, prostate herbs. Leaves small, opposite, ovate-lanceolate, or subspathulate, 3–6 × 1.5–4 mm, obtuse or subacute; petioles obscure; stipules scarious, minute. Flowers solitary in forks of branches, 2–2.5 mm long, subsessile. Corolla lobes ovate, acute, with a tooth on each side. Capsule obovoid oblong, 2–4 mm long, crowned with persistent calyx, densely covered with coarse hairs, seeds minute, angular, black.

Fls. & Frts.: April–October. *Exsiccata*: SUK 2284

Common, marsh loving.

Occurrence: Malwan, Ratnagiri, Vaibhavwadi.

Ecological Note: It grows on drying mud along margins of ponds and streams.

Oldenlandia corymbosa L., Sp. P1. 119. 1753; Hook. f. Fl. Brit. India 3: 65. 1880, Cook, in Aquat. & Wetl. Pl. India 337. 1996.

Annual herbs. Leaves subsessile, 2–4.5 × 0.3–0.4 cm, linear or linear-lanceolate, acute, often recurved and frequently scabrous margin. Flowers on filifrom pedicels longer than calyx, usually 2–3 on the top of a very slender axillary solitary pedundes; bracts beneath the pedicel, long subulate. Capsule globose or sometimes slightly pyrifrom somewhat didymous, the top rather flat and not protruded beyond the calyx, glabrous. Seeds pale brown, angular.

Fls. & Frts.: September–November. *Exsiccata*: SUK 2338

Common, marsh loving.

Occurrence: Malwan, Ratnagiri, Vaibhavwadi.

Ecological Note: It grows on drying mud along margins of ponds and streams.

Oldenlandia diffusa (Willd.)Roxb. Fl. Ind. 1: 444. 1820; Hook. f. Fl. Brit. India 3: 64. 1880. *Hedyotis diffusa* Willd., Sp. P1. 566. 1797

Annual, diffuse, glabrous herbs. Leaves subsessile, 2–3.5 × 0.2–0.3 cm, linear or linear-lanceolate, acute, glabrous; stipules short, slightly pectinate. Flowers usually solitary, sessile or on very short pedicels. Capsules broader than long, 2–2.5 mm, very truncate and flat on top. Seeds angular.

Fls. & Frts.: August–October. *Exsiccata*: SUK 2374

Common, marsh loving.

Occurrence: Vaibhavwadi

Ecological Note: It grows in open grassland.

SCROPHULARIACEAE

Bacopa monnieri (L.) Wettst in Engl. & Prantl., Pflaneerfam. 4 (36): 77. 1891; Pennell. In Proc. Acad. Nat. Sci. Phalad. 68: 94. 1946. *Lysimachia monnieri* L. Mant. Pl. 2: 9. 1756. *Herpestis nibuera* (L.) Benth Scroph. Ind. 30. 1835; Hook f. Fl. Brit. India 4: 272. 1884.

Glabrous, succulent creeping herbs. Leaves sessile, decussate, 0.6–2.5 × 2–cm, obovate-oblong or spathulate; fleshy, dotted with red speets, obtuse, entire flowers solitary, axillary. Corolla pale blue or white, 0.8 cm long; lobes nearly equal or rounded. Anthers bluish purple; pollen white. Capsule 0.5 cm long, ovoid, acute, pointed with style base, glabrous.

Fls. & Frts.: January–June. *Exsiccata*: SUK 2340

Common, marsh plant.

Occurrence: Achirne, Bhatkal, Karwar, Lolium, Malwan, Maxem, Panjim, Ratnagiri.

Ecological Note: It grows in ditches, margins of lakes and in muds.

Centranthera nepalensis. D. Don. Prodr. 88. 1825; Pennell, Acad, Nat. Sci. Philad. Monogr.5: 93. 1943. *C. hispida* auct. Plur. non R. Br.; Hook. f. Fl. Brit. India, 4: 301. 1884. *Rhinacanthus indica* L., Sp. Pl. 663. 1753.

An erect, hispid often hoary herbs. Leaves variable 1.2–3 × 0.3–1 cm, subsessile, linear-oblong or sometimes elliptic-oblong; hairy. Flowers rather large, subsessile in bracteate spike. Corolla infundibuliform rose-purple or rarely white, 2 cm long; lobes subequal, rounded filaments sparsely having, anther cells spurred, one perfect one sterile. Capsules 0.6–1cm long, broadly ovoid-oblong, grooved, slightly rugose.

Fls. & Frts.: September–February. *Exsiccata*: SUK 2340

Common, marsh loving.

Occurrence: Vaibhavwadi

Ecological Note: It grows in wet places on grasslands.

Dopatrium junceum (Roxb.) Buch.-Ham. ex Benth., Scroph. Ind., 31. 1835; Hook. f. Fl. Brit. India 4: 274. 1884. *Gartiola juncea* Roxb. Fl. Corom. 2: 16 Pl. 129. 1800.

Erect, annual herb; stems acutely 4-angled, glabrous. Leaves opposite, sessile, the lower oblong-elliptic or oblanceolate, 10–25 × 2.5 mm, obtuse, the upper much smaller and passing into bracts, Flowers axillary, solitary; pedicels 0. 2–15 cm long, slender. Corolla bluish-purple, 4–5 mm long, obscurely 2-lipped lobes. Capsule globose or ovoid, 2 mm long, 4-valved. Seeds numerous, minute, oblong, tuberculate.

Fls. & Frts.: August-February. *Exsiccata*: SUK 2342

Common, marsh loving.

Occurrence: Ratnagiri, Vaibhavwadi.

Ecological Note: It grows in temporary ponds, puddles and in water logged rice fields.

Glossostigma diandrum (L.) O. Ktze. Rev. Gen. Pl. 461. 1981. *Limasella diandra* L., Mant. Pl. 252. 1767. *Glossostigma spathulatum* (Hook.) Arn. ex Benth in Comp. Bot. Mag. E: 59. 1836; Hook. f. Fl. Brit. India 4: 28. 1884.

Tiny, tufted, creeping herb. Leaves linear-spathulate or oblanceolate, 2–8 × 1–2 mm, narrowed into a flat petiole, obtuse, glabrous, flowers axillary, solitary; pedicels slender, 4–12 mm long. Corolla pale blue pink or whitish; tubes included in the calyx; limb 2-lipped; lobes ovate-subacute. Stamens 2.

Fls. & Frts.: December–February *Exsiccata*: SUK 2353

Occasional, marsh loving.

Occurrence: Kolhapur, Karwar.

Ecological Note: It grows in wet muddy places along margins of lakes.

Limnophila aromatica (Lam.) Merrill, Interp. Herb. Amb. 466, 1917 & Enum. Philipp. Fl. Pl. 3: 432.1923. *Ambulia aromatica* Lam. Encycl. 1: 128.1783. *Lindernia gratissima* Blume, Bijdr. 749. 1826.

Stems erect, simple or branched from the base, glabrous to minutely glandular. Leaves decussate or in whorls of 3, simple; blade sessile, ovate-lanceolate to lanceolate-elliptical. Flowers solitary axillary or in many-flowered, terminal and axillary racemes; pedicels slender. Petals 10–18 mm long, pink or mauve or purple with yellow in throat, villous inside. Capsules broadly ellipsoid, 4–6 mm long. Seeds cylindrical to oblong, muricate.

Fls. & Frts.: November–April. *Exsiccata*: SUK 2289

Rare, marsh loving.

Occurrence: Amba, Dharwad, Karwar.

Ecological Note: It grows in harvested rice fields and on margins of pools.

Lindernia anagallis (Burm. f.) Pennell in J. Arnold. Arbor. 24. 252. 1943. *Ruellia anagallis* Burm. f. Fl. Ind. 135. 1768. *Vandellia pedunculata* Benth. Scrop. India 37. 1835; Hook. f. in Fl. Brit. India 4: 282. 1884.

Small glabrous diffusely branched herb. Leaves sessile or subsessile, 1.2–2 × 0.8–1.2 cm long, ovate, subacute, crenate-serrate, glabrous, base subcordate. Flowers solitary, axillary; pedicels 2–3 cm long. Corolla 0.8–1 cm long, white with pale violet margin, the lower lip marked with a yellow spot. Capsules twice as long as calyx, linear-lanceolate, acute, tipped by persistent style base.

Fls. & Frts.: August–December. *Exsiccata*: SUK 2288

Common, emergent hydrophyte.

Occurrence: Vaibhavwadi

Ecological Note: It grows in muddy places.

Lindernia angustifolia (Benth.) Wettst. In Engl. & Prantl. Pflanzenfam. 4(3b): 79. 1981; Mukerjee in J. Ind. Bot. Soc. 24: 132. 1945. *Vandelia angustifolia* Benth. Scrop. Ind. 57. 1835; Hook. f. Fl. Brit. India 4: 282–3. 1884.

A prostrate creeping herb, rooting at nodes. Flowers in terminal and axillary racemes, also axillary solitary, bracteate, pedicellate, 2- lipped; pedicel 1 cm long; leafy. Corolla ca 1.1 cm long, cylindrical, yellowish; upper lip outermost in bud, entire; lower lip 3-lobed, Stamens 2, perfect, arising from throat; filaments 2 mm long; anthers white, lobes divergent; staminodes 2, attached to the middle lobe of lower lip, yellow, clavate, papillose, arising from throat. Capsules ca 1.5 cm long, exceeding the calyx, ellipsoid, beaked.

Fls. & Frts.: July–January. *Exsiccata*: SUK 2381

Common, marsh loving.

Occurrence: Mapusa, Malwan, Punjim, Ratnagiri,, Shiroda, Udupi.

Ecological Note: It grows in marshy or wet situations such as paddy fields.

Lindernia antipoda (L.) Alston in Trim. Handb. Fl. Ceyl 6 (Suppl. 214. 1931). Pennell in J. Arn. Arb. 20: 81. 1939. *Ruellia antipoda* L., Sp. Pl. 635. 1753. *Bonnaya veronicifolia* (Retz.) Spreng. Syst. 1: 41. 1821; Hook. f. Fl. Brit. India 4: 285. 1884.

A prostrate creeping herb. Flowers in terminal and axillary racemes, also axillary, solitary, bracteate, pedicellate, 2-lipped; bracts leafy. Corolla yellowish; upper lip outermost in bud, entire; lower lip 3-lobed, Stamens 2, perfect, arising from throat; filaments 2 mm long; staminodes 2, attached to the middle lobe of lower lip, yellow, clavate, papillose, arising from throat. Capsules exceeding the calyx, ellipsoid, beaked. Seeds yellow, pointed at both ends.

Fls. & Frts.: July–January. *Exsiccata*: SUK 274

Common, marsh loving.

Occurrence: Mapusa, Malwan, Ratnagiri, Vaibhavwadi.

Ecological Note: It grows in marshy or wet situations such as paddy fields.

Lindernia ciliata (Colsm.) Pennell in Brittonia 2: 182. 1936; Sant. in Rec. Bull. Bot. Surv. India 16 (1): 180. ed. 3. 1967. *Gratiola ciliata* Colsm. Desc. Grat. 14. 1793. *Bormaxa brachiata* Link. & Otto. Ic. Pl. Select. 25. t. 11. 1820, Hook. f. Fl. Brit. India 4: 284. 1884.

An erect or prostrate herb growing upto 15 cm long. Leaves 1–3 × 0.5–1 cm, opposite, sessile, elliptic-oblong, clasping the the stem at base, distantly serrate at margins, mucronulate at apex, pinnately nerved. Flowers in terminal and axillary racemes, bracteate, almost sessile, 2-lipped; bracts 3 mm long, linear. Corolla ca 1 cm long, cylindrical, pink to violet in colour; tube 4 mm long, hairy; upper lip outermost in bud, notched at apex; lower lip 3-lobed, Stamens 2, perfect, arising from throat; staminodes 2. Capsules ca 1 cm long, exceeding the calyx, ellipsoid, purple in colour.

Fls. & Frts.: July–September. *Exsiccata*: SUK 2341

Abundant, marsh-land plant.

Occurrence: Malwan, Ratnagiri, Shiroda, Vaibhavwadi.

Ecological Note: It grows in marshy or wet situations such as paddy fields.

Lindernia crustacea (L.) F. Muell. Syst. Census. Austr. L. 1: 97. 1882. *Caprania crustacea* L., Mant. Pl. 87. 1767. *Vandellia crustacea* (L.) Benth. Scroph. Ind. 35. 1835; Hook. f. Fl. Brit. India 4: 279. 1884.

An erect or prostrate creeping herb. Leaves exstipulate, petiolate, ovate to elliptic, truncate or attenuate at base, entire serrate or crenate along margins, acute at apex, pinnately nerved, nerves purple in colour. Flowers in axillary solitary and in terminal racemes, ebracteate, pedicellate, 2-lipped. Corolla ca 7 mm long, purple; tubes cylindrical; upper lip outermost in bud, 2-lobed; lower lip 3-lobed, yellow spot on the throat. Stamens 4. Ovules many. Capsules ca 4 mm long, equaling the calyx, ovoid.

Fls. & Frts.: July–November. *Exsiccata*: SUK 2292

Common, marsh loving.

Occurrence: Malwan, Vaibhavwadi.

Ecological Note: It grows on lateritic rocks, marshy or wet situations and field bunds.

Lindernia estaminoidiosa (Blatt. & Hallb.) Mukerjee in J. Ind. Bot. Soc. 24: 133. 1945; Sant. in J. Bombay nat. Hist. Soc. 49: 40. 1950. *Bonnaya estaminodiosa* Blatt. & Hallb. in J. Bombay nat. Hist. Soc. 25: 416. 1918.

A prostrate creeping herb growing upto 17 cm long, rooting at nodes. Leaves 0.5–2.5 × 0.3–0.8 cm, opposite decussate, exstipulate, petiolate, elliptic-oblanceolate, attenuate at base, distantly serrate, acute at apex, pinnately nerved; petioles 0.5 cm long. Flowers in terminal and axillary racemes, also axillary solitary, bracteate, pedicellate, 2-lipped; pedicel 1 cm long. Corolla ca 1.1 cm long, cylindrical, yellowish; upper lip outermost in bud, entire; lower lip 3-lobed, Stamens 2, perfect; staminodes 2, Capsules ca 1.5 cm long, exceeding the calyx, ellipsoid, beaked.

Fls. & Frts.: July–January. *Exsiccata*: SUK 2344

Frequent, marsh loving.

Occurrence: Mapusa, Ratnagiri. Endemic.

Lindernia hyssopioides (L.) Haines, Bot. Bihar & Orissa 635. 1922. *Gratiola hyssopioides* L., Mant. Pl. 174. 1771. *Ilysanthes hyssopioides* (L.) Benth. in DC. Prodr. 10: 419. 1846; Hook. f. Fl. Brit. India 4: 283. 1884.

An erect herb growing upto 30 cm high. Flowers axillary, solitary, ebracteate, pedicellate, 2-lipped; pedicel 3 mm long; deflexed in fruits. Corolla ca 1.5 cm long, violet; upper lip outermost in bud, 2-lobed. with violet strips; lower lip 3-lobed, Stamens 2; staminodes 2. Capsules ca 8 mm long, exceeding the calyx, ellipsoid, beaked.

Fls. & Frts.: August–January. *Exsiccata*: SUK 2373

Common, marsh loving.

Occurrence: Malwan, Vaibhavwadi.

Ecological Note: It grows in marshy or wet situations such as paddy fields.

Lindernia nummularifolia (D. Don) Wettst. In Engl. & Prantl. Pflanzenfam 4 (3b.): 79. 1891. *Vandelia nummularifolia* D. Don Prodr. Fl. Nepal 86. 1825. *Vandelia sessiliflora* Benth. Scroph. Ind. 37. 1835; Hook. f. Fl. Brit. India 4: 282. 1884.

Small herb. Leaves shortly petioled; blades suborbicular to ovate-orbicular, 1–2.5 cm across, glabrous above, sparsely hairy beneath, base truncate, margins serrate, ciliolate, tip obtuse. Flowers solitary, axillary, or in lax racemes. Petals 7–10 mm long, white and pink or white to purple with a yellow fleck at the throat. Stamens 4. Capsules linear-cylindrical, acuminate, 5–11 mm long, the wall almost transparent.

Fls. & Frts.: August–April. *Exsiccata*: SUK 2294

Common, marsh loving.

Occurrence: Malwan, Vaibhavwadi.

Ecological Note: It grows in marshy places.

Lindernia oppositifolia (Retz.) Mukerjee in J. Ind. Bot. Soc. 24: 134. 1945; Shah Fl. Gujrath 502. 1976. *Gratiola oppositifolia* Retz. Obs. 4: 8. 1786. *Bonnaya oppositifolia* Spreng. Syst. Reg. 1: 41. 1825; Hook. f. Fl. Brit. India 4: 286. 1884.

A prostrate creeping herb. Leaves 0.5–2.5 × 0.3–0.8 cm, opposite decussate, exstipulate, petiolate, elliptic-oblanceolate, attenuate at base, distantly serrate, acute at apex. Flowers in terminal and axillary racemes, also axillary solitary, bracteate, pedicellate, 2-lipped; pedicel 1 cm long; bracts leafy. Corolla ca 1.1 cm long, cylindrical, yellowish; upper lip outermost in bud, entire; lower lip 3-lobed, Stamens 2, perfect; staminodes 2. Capsules ca 1.5 cm long, exceeding the calyx, ellipsoid, beaked.

Fls. & Frts.: July–January. *Exsiccata*: SUK 2277

Frequent, marsh loving.

Occurrence: Udupi, Vaibhavwadi.

Lindernia parviflora (Roxb.) Haines, Bot. Bihar. & Orissa 635. 1922 *Gratiola parviflora* Roxb. Pl. Corom. 3: 3, t. 203. 1811. *Ilysanthes parviflora* (Roxb.) Benth. in DC. Prodr. 10. 419. 1846; Hook. f. Fl. Brit. India. 4: 283. 1884.

An erect or prostate herb up to 20 cm long. Leaves sessile, elliptic, cuneate at base, entire, acute at the apex, 3-nerved from the base. Flowers axillary, solitary, ebracteate, pedicellate, 2-lipped; pedicel extending up to 3 cm long and deflexed in fruit. Corolla ca 7 mm long, white or blue tube ca 3 mm long, cylindrical; upper lip outermost in bud, 2-lobed; lower lip 3-lobed, middle lobe with two blue spots. Stamens 2; staminodes 2, ca 4 mm long. Capsule ca 5 mm long, exceeding the calyx, ellipsoid-oblong, beaked.

Fls. & Frts.: Auguast–January. *Exsiccata*: SUK 2354

Common, marsh loving.

Occurrence: Alibag.

Ecological Note: It grows mostly in shallow ponds, marshy places and in paddy fields

Lindernia pusilla (Willd.) Bold. Zakhf. Landbowstr. Java. 165. 1956. *Gratiola pussilla* Willd., Sp. Pl. 1: 105. 1797. *Vandellia scabra* Benth. Scroph. Ind. 36. 1835; Hook. f. Fl. Brit. India 4: 281. 1884.

Prostrate, ascending or erect herb. Leavers sessile or subsessile, ovate to suborbicular, 5–12x4–14 mm, pubescent above, hirsute beneath, base truncate, margins obscurely toothed, tips acute. Flowers solitary or in clusters of 4 in axils. Petals 5–8 mm long, violet; yellow or white at throat, usually with a brown adaxial lip. Stamens 4, all with fertile anthers. Capsules obovoid or globose.

Fls. & Frts.: August–December. *Exsiccata*: SUK 2296

Frequent, marsh loving.

Occurrence: Malvan, Roha.

Ecological Note: It grows mostly in shallow ponds, marshy places and in paddy fields

Lindernia rotundifolia (L.) Alston in Trimen, Handb. Fl. Ceylon 6: 214. 1931; Mukerjee in J. Ind. Bot. Soc. 24: 132. 1945. *Gratiola rotundifolia* L. Mant. 174. 1771. *Ilysanthes rotundifolia* (L.) Benth. in DC. Prodr. 10: 420. 1846; Hook. f. Fl. Brit. India 4: 284. 1884.

An erect or prostrate herb, rooting at the lower nodes. Leaves opposite, sub-orbicular, entire, acute at apex, 3-nerved from base; pedicel 5 mm long. Corolla 4–5 mm long, white or blue; tube 2 mm long; upper lip 2-lobed, uppermost in bud; lower lip 3-lobed. Stamens 2, perfect, arising from throat; staminodes 2, elongated, glandular, yellow at base, arising from throat. Capsules upto 3 mm long, equaling the calyx, globose with persistent style.

Fls. & Frts.: August–January. *Exsiccata*: SUK 2275

Common, marsh loving.

Occurrence: Bhatkal.

Ecological Note: It grows in marshy and moist areas.

Lindernia tenuifolia (Cols.)Alston in Trim. Fl. Ceylon 6. Suppl. 214. 1935 *Grantiola tenuifolia* Colsm. Prodr. Desc. Grat. 8. 1793. *Bonnaya tenuifolia* (Colsm.) Spren. Syst. 1: 42. 1842; Hook. f. Fl. Brit. India 4: 286. 1884.

A creeping herb, rooting at nodes. Leaves opposite decussate, exstipulate, petiolate, elliptic-oblanceolate, attenuate at base, distantly serrate, acute at apex, pinnately nerved. Flowers in terminal and axillary racemes, also axillary solitary, bracteate, pedicellate, 2-lipped; pedicel 1 cm long; bracts leafy. Corolla yellowish; upper lip outermost in bud, entire; lower lip 3-lobed, Stamens 2; staminodes 2. Capsules ca 1.5 cm long, exceeding the calyx, ellipsoid, beaked.

Fls. & Frts.: July–January. *Exsiccata*: SUK 2213

Common, marsh loving.

Occurrence: Mapusa, Malwan.

Ecological Note: It grows in marshy or wet situations such as paddy fields.

Lindernia viscosa (Hornem.) Boldingh, Zakfl. Handb. Java. 165, 1916; Sant. in Fl. Khandala (ed.) 3 179, 1967. *Gratiola viscosa* Hornem. Pl.

Hort. Hafn; 19, 1807. *Vandelia hirsuta* Buch.-Ham. ex Benth. Scroph. Ind. 36, 1835; Hook. f. Fl. Brit. India 4: 280. 1884.

A prostrate herbs. Leaves elliptic-oblanceolate, attenuate at base, distantly serrate, acute at apex, pinnately nerved. Flowers in terminal and axillary racemes, also axillary solitary, bracteate, pedicellate, 2- lipped; pedicel 1 cm long; bracts leafy. Corolla ca 1.1 cm long, cylindrical, yellowish; upper lip outermost in bud, entire; lower lip 3-lobed, Stamens 2, perfect, arising from throat; filaments 2 mm long; anthers white, lobes divergent; staminodes 2, attached to the middle lobe of lower lip, yellow, clavate, papillose, arising from throat. Capsules ca 1.5 cm long, ellipsoid, beaked.

Fls. & Frts.: July–January. *Exsiccata*: SUK 2215

Common, marsh loving.

Occurrence: Mapusa, Malwan,

Ecological Note: It grows in marshy or wet situations such as paddy fields.

Ramphicarpa longiflora (Arn.) Benth. Compd. Bot. Mag. 1: 368. 1836; Hook. f. Fl. Brit. India 4: 377. 1958 (Repr. ed). Cook in Aquat & Wetl. Pl. India 362. 1996.

Erect, branched, glabrous herb. Lower leaves opposite, the upper alternate, 2.5–3.5 cm long, pinnatisect with filiform segments. Flowers axillary, shortly pedicelled, or in branched racemes; bracteoles l or 2 or rarely absent. Corolla white; tube elongate, very slender, straight, 2.5–3.5 cm long; limb 1.2–2 cm across, lobes 5, spreading, obovate-oblong, obtuse, emarginate. Capsules 1. 2 cm long, ovoid, compressed at angles to Septum, beaked.

Fls. & Frts.: August–February. *Exsiccata*: SUK 1984

Common, marshy loving.

Occurrence: Lolium, Vaibhavwadi.

Ecological Note: It grows on lateritic plateaus as well as in grasslands.

VERBENACEAE

Phyla nodiflora (L) Greene in Pittonia 4: 46. *Verbena nodiflora* L., Sp. Pl. 20. 1753. *Lippia nodiflora* A. Rich. in Michx. Fl. Bor. Amer. 2: 15. 1803; Cl. in Hook. f. Fl. Brit. India 4: 563. 1885. GOUR MUNDI

Prostrate, perennial herbs. Leaves opposite, subsessile, obovate or spathulate, 1.5–3 × 1–2.5 cm, cuneate at base, deeply serrate, obtuse or rounded at apex, appressed pubescent on both surfaces. Flowers minute, sessile, in densely packed, subglobose to oblong, 1.5–2 cm long head; peduncles 5–7 cm long; bracts obovate, mucronate. Fruits oblong, 1–1.5 mm long, flat, splitting, into 2, one-seeded pyrenes.

Fls. & Frts.: September–April. *Exsiccata:* SUK 2201

Common, marsh loving.

Occurrence: Kolhapur.

Ecological Note: It grows on margins of permanent tanks and lakes.

Monocots

ARACEAE

Cryptocoryne retrospiralis (Roxb.) Kunth, Enum. Pl. 3: 12. 1841; Hook. f. Fl. Brit. India 6: 493. 1893. Subr. Aquat. Ang. 76. 1962; Cook. Aquat. Wetl. Pl. Ind. 57. 1996. *Ambrosinia retrospiralis* Roxb. Fl. Ind. 3: 492. 1832.

Herbs, roots fleshy, fibrous. Leaves radical, 7.5–45 × 0.6 cm, sessile or with a short stout petiole, midrib slender. Spathes subsessile, nearly as long as leaves, deep green, streaked with purple. Male and female flowers separated by naked rachis. Fruit 1.2 × 0.8 cm, ovoid, on a solitary stalk. Seeds 2-seriate, oblong, subtrigonous.

Fls. & Frts.: October–February. *Exsiccata*: 6077

Occasional, marsh loving.

Occurrence: Gaganbowda.

Ecological Note: It grows on wet margins of stream, river and lakes.

Cryptocoryne spiralis (Retz.) Fischer ex Wilder in Linnaea 5: 428. 1830; Hook, f. Fl. Brit. Ind. 6: 494. 1893; Cook. Aquat. Wetl. Pl. Ind. 57. 1996. *Arum spirale* Retz. Os. Bot. 1: 30. 1790. *Cryptocoryne huegeli* Schott, Aroid. 8. t. 12. 1853. *C. tortuosa* Blatt. & McC. in J. Bomb. Nat. Hist. Soc. 35: 15. t. 1. 1931.

Herbs; rootstock tuberous, stoloniferous; root vermiform. Leaves linear-lanceolate, acute or acuminate, narrowed from middle to both ends; Spathe subsessile, linear-lanceolate, green outside, dark purple and transverse lamellate within. Fruit subglobose. Seeds spindle shaped, irregularly verrucose-regose.

Fls. & Frts.: August–December. *Exsiccata*: 36

Common, marsh loving.

Occurrence: Vaibhavwadi, Ratnagiri, Udupi.

Ecological Note: It grows on marshy margin of ponds, pools, ditches and streams.

Cryptocoryne spiralis var. cognatoides (Blatt. & McC.) Yadav, Patil and Bogner in Aqua-Planta 2: 65. 1993. C. cognatoides Blatt. & Mcc. in J. Bomb. Nat. Hist. Soc. 35: 17. 1931.

Herbs, rhizomatous; root thick, cylindrical. Leaves ca 8 cm long, undulate; petiole upto 15 cm long. Spathes upto 14 cm long, linear-lanceolate; spadix 4 cm long. Male flowers in cone like clump at the top and female flowers grouped at base, separated by 2.5 cm long interspace.

Fls.: August–September. *Exsiccata*: 6076

Rare, marsh loving.

Occurrence: Amboli. Endemic.

Ecological Note: It grows in muddy places at higher altitude. (1000–1300m)

BURMANNIACEAE

Burmannia pusilla (Wall. ex Miers) Thw. Enum. Pl. Zeyl. 325. 1864; Hook. f. Brit. Ind. 5: 665. 1888; Cook, Aquat. Wetl. Pl. Ind. 80. 1996. *Tripteranthus pusilla* Wall. ex Miers in Trans. Linn. Soc. 18: 537. 1841. *Burmannia coelestis* var. *pusilla* (Wall. ex Miers) Trin. Handb. Fl. Ceylon 4: 131. 1898.

Stem slender, simple or branched, green. Leaves very few, scale like, sessile along the stem, lanceolate-subulate. Flowers obcordate or orbicular in outline, solitary or 2–3 together. Calyx tube winged; wings rounded, truncate or retuse at top. Petals minute.

Fls. & Frts.: August–September. *Exsiccata*: 443

Common, marsh loving.

Occurrence: Vaibhavwadi, Lolium, Manipal.

Ecological Note: It grows on lateritic plateaus.

COMMELINACEAE

Commelina benghalensis L. Sp. Pl. 41. 1758; Hook. F. Brit. India 6: 370. 1892; Cook, Aquat. Ang. 86. 1996. KENA

A branched herb, 2–3 feet in height. Stem subscandent, slender, dichotomously branched. Leaves far apart from the stem, ovate or oblong, obtuse, shortly petiolate, pubescent or villous on the both surfaces. Flowers both chasmogamous and cleistogamous. Capsule pyriform

Fls. & Frts.: June–December. *Exsiccata*: 1950

Common, marsh loving.

Occurrence: Throughout coastal plains.

Ecological Note: It grows in wetlands and along the streams, ponds and canals.

Commelina diffusa Burm. f. Fl. Brit. Ind. 18, t. 7, f. 2. 1768; Cook, Aquat. Wetl. Pl. Ind. 86. 1996. *C. nudiflora* sensu Hook. f. Fl. Brit. India 6: 369. 1892. GANDOLGI

Annual, diffuse, nearly glabrous herbs. Stem 1–3 feet, upwards branching from the base, rooting at nodes. Leaves sessile lanceolate, ovate-lanceolate, acute or acuminate, glabrous or puberulous. Cymes usually two, 1–3 flowered. Flowers blue; interior tepals obovate with long claws; exterior subsessile, orbicular, sometimes obsolete; ovary 3-celled. Capsules broadly oblong, coriaceous.

Fls. & Frts.: July–February. *Exsiccata*: 486

Common, marsh loving.

Occurrence: Throughout coastal plains.

Ecological Note: It grows in wet open land and near the water sources.

Commelina hasskarli Cl. Commel. & Cyrt. Beng. 13, t. 3. 187 & In DC. Monogr. Phan. 3: 157. 1881; Hook. f. Fl. Brit. Ind. 6: 370. 1892; Cook, Aquat. Wetl. Pl. Ind. 86. 1994.

Herb, decumbent, much branched, glabrous or pubescent. Leaves narrowly lanceolate, subacute; sheaths with broad base, ciliate. Spathes axillary, scattered, ovate-lanceolate, cordate at the base with rounded lobes. Flowers in pubescent, 1–2 flowered cymes. Capsule quadrate, subtruncate, apiculate, membranous. Seeds cylindric, truncate at one end, subacute or rounded at other, without an appendages.

Fls. & Frts.: June–March. *Exsiccata*: 1395

Common, marsh loving.

Ecological Note: It grows near the water sources such as stream, ponds, puddles etc.

Commelina subulata Roth, Nov. Pl. Sp. 23. 1821; Hook. f. Fl. Brit. Ind. 6: 369. 1892; Cook, Aquat. Wetl. Pl. Ind. 87. 1996.

Stem erect, slender, simple or branched, nearly glabrous. Leaves 3–7 cm long, linear, glabrous; sheaths 2 cm long, with glabrous or ciliate margin. Spathes fulcate, pubescent, green-purple, base cordate. Flowers orange-purple, when dry violet. Capsules small, ovoid, 5-seeded. Seeds black, puberulous, slightly compressed, deeply transversely grooved.

Fls. & Frts.: July–October. *Exsiccata*: 343

Common, marsh loving.

Occurrence: Amboli, Vaibhavwadi.

Ecological Note: It grows near water sources.

Flascopa scandens Lour. Fl. Cochinch. 193. 1790; Hook. f. Fl. Brit. Ind. 6: 390. 1893; Cook, Aquat. Wetl. Pl. Ind. 88. 1994.

Herbs, erect or subscandent. Leaves alternate, elliptic-lanceolate, much narrowed at the base, subsessile or shortly petiolate, scaberulous above, glabrous or pubescent beneath. Flowers in terminal, sessile or shortly pedunculate, villous or hirsute, clustered panicle. Capsules rather broader than long, suborbicular, abruptly subacute at the ends, shining. Seeds pale glaucous, dorsally transversely wrinkled.

Fls. & Frts.: September–October. *Exsiccata*: 1020

Rare, marsh loving

Occurrence: Amboli.

Ecological Note: It grows near the water sources.

Murdania nudiflora (L.) Brenan in Kew Bull. 189. 1952; Cook, Aquat. Wetl. Pl. Ind. 90. 1994. *Commelina nudiflora* L. Mant. 177. 1767 (non Sp. Pl. 41. 1753). *Aneilema nudiflorum* (L.) R. Br. Prodr. 271. 1810; Hook. f. Fl. Brit. India 6: 378. 1893.

Annual, erect or suberect herbs. Leaves linear, oblong or lanceolate, acute or acuminate, flat, glabrous, base rounded, cordate or narrowed into sheath; sheaths hairy. Flowers in terminal panicles, branches of panicles long, scarred above; flowers in subglobose or subcorymbose cymes at the ends of the branches. Capsules trigonously subglobose, membranous, mucronate, 3-celled with 2 seeds in each cell. Seeds truncate at one end, rounded at other.

Fls. & Frts.: July–November. *Exsiccata*: 1568

Common, marsh loving.

Occurrence: Ratnagiri, Udupi.

Ecological Note: It grows in wetland.

Murdania pauciflora (Wright) Brueck, in Engl. Prantl. Pflanzen. Fam. ed. 2, 15a. 173. 1930; Cook, Aquat. Wetl. Pl. Ind. 90. 1994; *Aneilema pauciflorum* Wight, Ic. t. 2077. 1853 non Dalz. 1851; Hook. f. Fl. Brit. Ind. 6: 378. 1893.

Stems creeping and ascending, much branched; rooting at nodes. Leaves sessile, ovate, subacute, hairy, base rounded or cordate; sheaths with ciliate margins. Flowers 1–3, in axillary cymes. Capsules oblong-fusiform, acute, somewhat 3-gonous, smooth, 3-celled. Seeds 4–6 in each cell, 1-seriate, superposed, cubical, nearly smooth.

Fls. & Frts.: July–November. *Exsiccata*: 5297

Common, marsh loving.

Occurrence: Dastikoppa. Endemic.

Ecological Note: It grows near the marshy sources.

Murdania semeteres (Dalz.) Sant. in Poona Agric. Coll. Mag. 41: 284. 1951 and J. Bomb. Nat. Hist. Soc. 52: 658. 1954; Cook, Aquat. Wetl. Pl. Ind. 90. 1996. *Aneilema semeteres* Dalz. J. Bot. 3: 138. 1851. *A. paniculatum* Wall. ex Clarke, in DC. Monogr. Phan 3: 215. 1881 non Wight, 1853; Hook f. Fl. Brit. India 6: 381. 1892.

Erect small herbs; root fibrous; stem sheathed at the base with yellowish scarious sheath. Leaves linear, finely acuminate. Flowers in paniculate cymes; bracts ochreate, cyathiform, truncate, with a teeth at apex, persistent. Capsules ellipsoid or subglobose, smooth. shining. Seeds 6–8 in each cell, pale yellow, 2-seriate, irregularly angular, smooth.

Fls. & Frts.: July–October. *Exsiccata*: 431

Common, marsh loving.

Occurrence: Honnawar, Malwan, Udupi.

Ecological Note: It grows near the water sources.

Murdania spirata (L.) Brueck. in Engl. Prantl.. Pflanzenfam. ed. 2, 15a: 173. 1930; Cook, Aquat. Wetl. Pl. Ind. 91. 1996. *Aneilema spiratum* (L.) R. Br. Prodr. 271. 1810; Hook. f. Fl. Brit. India 6: 377. 1892.

Herbs. Leaves sessile; lower broadly linear-lanceolate; upper oblong, cordate, all amplexicaul, glabrous, acute, canaliculated in the middle; sheaths very short. Inflorescence dichotomously branched, panicle, few flowered. Flowers bluish-purple or pale white. Capsules oblong, mucronate, smooth, trigonous, 3–8 seeds in each cell. Seeds subconical-cylindrical.

Fls. & Frts.: July–March. *Exsiccata*: 153

Common, marsh loving.

Occurrence: Udupi

Ecological Note: It grows near the water sources.

Murdania vaginata (L.) Brueck. in Engl. & Prantl. Pflazenfam. ed. 2, 15a: 173. 1930; Sant. J. Bomb. Nat. Hist. Soc. 52: 658. 1954; Cook, Aquat. Wetl. Pl. Ind. 91. 1996. *Commelina vaginata* L. Mant. 177. 1767. *Aneilema vaginatum* (L.) R. Br. Prodr. 271. 1810; Hook. F. Fl. Brit. Ind. 6: 381. 1893.

Herbs; rooting at the lower nodes; branching suberect, ending in filiform peduncles with distant flowering bracts. Leaves linear, acuminate, flat, glabrous or sparsely hairy; sheaths short, open, ciliate. Flowers 1–3 together, fascicled in distant bracts, along the filiform very slender terminations of the branches. Capsule globose, cuspidate, 3-celled; cells 1-seeded or empty. Seeds trigonous, truncate at one end other narrowed.

Fls. & Frts.: August–October. *Exsiccata*: 450

Occasional, marsh loving.

Occurrence: Kittur, Ratnagiri, Sawantwadi, Udupi.

Ecological Note: It grows in wet places and rice fields.

Murdania versicolor (Dalz.) Brueck. in Engl. & Prantl. Pflanzenfam. ed. 2, 15a. 173. 1930; Cook, Aquat. Wetl. Pl. Ind. 91. 1996. *Aneilema versicolour* Dalz. J. Bot. 3: 136. 1851; Hook. f. Fl. Brit. Ind. 6: 378. 1892.

Small herb with suberect branches. Leaves alternate, bifarious, 5–6 × 1.7 cm, flat, oblong-lanceolate, acuminate, stem clasping; sheaths furrowed, hispid. Flowers ochre-yellow; pedicels axillary, filiform, jointed about middle. Capsules scarcely longer than persistent sepals, linear, subtrigonous, pointed, erect; cells 5–7-seeded.

Fls. & Frts.: August–October. *Exsiccata*: 413

Occasional, marsh loving.

Occurrence: Amboli, Lolium, Malwan, Ratnagiri, Sawantwadi, Udupi. Endemic.

Ecological Note: It grows in wet fields as well as near water sources such as ponds, puddles etc.

CYPERACEAE

Bulbostylis densa (Wall.) Handel-Mazzetti in Karsten & Schenk, Vegetatiensh 20 (7): 16. 1930; Cook, Aqua. Wetl. Pl. Ind. 96. 1996. *Scripus densus* Wall. In Roxb. Fl. Ind. 1: 231. 1820. *Bulbostylis capillaris* Kunth var. *trifida* (Nees) Cl. in Hook. f. Fl. Brit. India 6: 652. 1893.

Tufted, annual herbs. Leaves glabrous; sheaths hairy. Spikelets few, solitary, elliptic, 3 mm across; glumes dark brown, broadly ovate, laxly imbricate. Nuts obovate, triquetrous, verruculose.

Fls. & Frts.: September-November. *Exsiccata*: 6089

Occasional, marsh loving.

Occurrence: Honnawar.

Ecological Note: Grows in wet and marshy land.

Cyperus alopecuroides Rottb. Descr. Pl. Rar. Programm. 20. 1772; Cook, Aquat. Wetl. Pl. Ind. 102. 1994.

Rhizome very woody. Leaves 3–5, sheaths somewhat spongy, yellowish-blackish brown; blade inserted, w-shaped, leathery. Inflorescence compound paniculate. Spike digitate upright, cylindrical to oblong-cylindrical. Glume ovate to ellipsoid, ovate, obtuse, apiculate, minutely mucronate.

Fls. & Frts.: September–January. *Exsiccata*: 6098

Category: Marsh loving.

Status: Common.

Occurrence: Honnawar.

Ecological Note: Grows in marshy land near the water sources.

Cyperus alulatus Kern. In Reinwardtia 1: 463, f. 1. 1952. Cook, Aqua. Wetl. Pl. Ind. 103. 1996. *C. iria* var. *rectangularis* Kuekenth in Engl. Pflanzens. 101. 152. 1935. *C. rectangularis* (Kuekenth) Bennet in Indian Forester 95: 692. 1969.

Herbs upto 30 cm in height. Leaves 16 cm, linear. Inflorescence terminal lax, compound umbels. Spikelets 4 × 2 mm, reddish or greenish-yellow. Nuts 1 mm long, obovoid, spiculate, brown.

Fls. & Frts.: July–September. *Exsiccata*: 6013

Common, marsh loving.

Occurrence: Malpe, Malwan, Mapusa, Panjim.

Ecological Note: It grows in wet and marshy land.

Cyperus articulatus L. Sp. Pl. 44. 1753; Clarke in Hook. f. Fl. Brit. Ind. 6: 611. 1893; Cook, Aquat. Wetl. Pl. Ind. 105. 1994.

Herbs, ca 3 m in height; root stock stout, stoloniferous. Umbel compound of many spikes; floral glumes 3–4 mm long. Nuts trigonous, oblong-ellipsoid, black.

Fls. & Frts.: September–December. *Exsiccata*: 1756

Common. marsh loving.

Occurrence: Mapusa, Malpe, Panjim, Pudubudri, Shiroda, Wanawar Ramthirth.

Ecological Note: It grows in wet and marshy lands.

Crperus castaneus Willd. Sp. Pl. 1: 278. 1779; Cl. in Hook. f. Fl. Brit. India. 6: 598. 1893; Cook, Aquat. Wetl. Pl. Ind. 105. 1994.

Small herbs; stem tufted. Leaves as long as or shorter than stem; sheaths often purple. Umbels simple, consisting central sessile heads and many unequal spreading rays; rays terminated by heads of 3–20 spikelets. Glumes greenish yellow with a recurved mucro. Nut oblong, obovoid, trigonous, apiculate, brown or yellow.

Fls. & Frts.: December–January. *Exsiccata*: 1378

Common, marsh loving.

Occurrence: Malpe, Malwan, Mapusa, Pudubudri.

Ecological Note: It grows in wet and marshy lands.

Cyperus compressus L. Sp. Pl. 46. 1753 emend Dandy in Exell. Cat. S. Tomes 357. 1944; Cl. in Hook. f. Fl. Brit. Ind. 6: 605. 1893; Cook, Aquat. Wetl. Pl. Ind. 106. 1996.

A glabrous, annual herb; stem tufted. Leaves basal, shorter or slightly longer than culms, linear, finely acuminate. Umbel simple. Spikelets much compressed, greenish yellow, linear oblong, 20–40 flowered. Glumes oblong-lanceolate, ciosely imbricate; keel produced into slightly recurved, laterally compressed mucro. Nuts broadly obovoid, ellipsoid, triquetrous, black, shortly beaked.

Fls. & Frts.: July–October. *Exsiccata*: 1726

Common, marsh loving.

Occurrence: Malpe, Mapusa, Roha, Pudubudri, Panjim.

Ecological Note: It grows near the water sources.

Cyperus corymbosus Rottb. Descr. Ic. 42, t. 7, f. 4. 1753; Cl. in Hook. f. Fl. Brit. India 6: 612. 1893; Cook, Aquat. Wetl Pl. Ind. 107. 1996. *C. tegetiformis* Roxb. ex. Arn. in Wight, Contrib. Bot. Ind. 89. 1834; Cl. in Hook. f. op. cit.

Herbs, perennial; rootstock tuberous. Stem erect, 1–2.5 m high, subtrigonous. Leaves linear. Inflorescence a decompound umbel. Spikelet alternate, 3 × 1 mm, greenish-brown, ovate. Nuts obovoid, apiculate.

Fls. & Frts.: August–November. *Exsiccata*: 178

Common, marsh loving.

Occurrence: Mapusa, Malpe, Malwan, Roha, Ratnagiri.

Ecological Note: It grows in wet and marshy land.

Cyperus cuspidatus Kunth in H.B. & K. Nov. Gen. Sp. 1: 204. 1815; Cl. in Hook. f. Fl. Brit. India 6: 598. 1893; Cook, Aquat. Wetl. Pl. Ind. 108. 1996. *C. uncinatus sensu*. Cl. in Philipp. J. Sci. 2: 82. 1907 *non* Poir. 1806.

Annual, glabrous, tufted herbs. Leaves as long as or shorter than stem, filiform, flexuous. Umbel simple, consisting central sessile and few to many spreading unequal rays. Spikelets in 5–12 cluster, compressed, 12–14 flowered. Glumes oblong-spathulate; keel nerved, mucro spreading or recurved. Nuts obovoid, trigonous, brown to castaneous.

Fls. & Frts.: October–November. *Exsiccata*: 42

Common, marsh loving.

Occurrence: Malpe, Panjim, Pudubudri, Sawantwadi.

Ecological Note: It grows in rice field, seasonally wet places and margin of ponds.

Cyperus difformis L. Sp. Pl. 67. 1753; Cl. in Hook, f. Fl. Brit. Ind. 6: 599. 1893; Cook, Aquat. Wetl. Pl. Ind. 109. 1996. LAVALA

Annual herbs. Leaves 10–15 cm long, linear. Inflorescence a terminal, simple or compound umbel with a sessile, capitate head at the base of each umbel. Spikelets brown. Nuts ovate-oblong, minutely punctate.

Fls. & Frts.: August–January. *Exsiccata*: 1054

Common, marsh loving.

Occurrence: Kolhapur, Malwan, Vaibhavwadi.

Ecological Note: It grows in wetlands, marshy places and in crop fields.

Cyperus distans L. f. Suppl. 103. 1781; Cl. in Hook. f. Fl. Brit. Ind. 6: 607. 1893; Cook, Aqua. Wetl. Pl. Ind. 110. 1994.

Perennial herbs; rhizome subterranean, stoloniferous; stems smooth, 50–55 cm tall. Leaves sub-basal, longer or equal to the stem, acuminate. Spikes in compound umbel; bracts 4–6, spreading, primary rays 6–12, upto 16 cm long; spikelets spicate, lower ones at right angle to rachis. Glumes oblong-ovate, obtuse. Nuts oblong-cylindrical or obovoid, blackish-brown.

Fls. & Frts.: September–November. *Exsiccata*: 41

Common, marsh loving.

Occurrence: Ratnagiri, Mapusa, Panjim, Pudbudri.

Ecological Note: It grows in wetlands and marshy places.

Cyperus exaltatus Retz. Obs. Bot. 5: 11. 1789; Cl. in Hook. f. Fl. Brit. Ind. 6: 617. 1893; Cook, Aquat. Wetl. Pl. Ind. 111. 1994.

A large glabrous herbs; stem 90–180 cm long, obtusely trigonous. Leaves few, linear, finely acuminate, margin scabrous. Umbel compound; rays 5–10, 15 cm long, spreading. Spikelets ascending, narrowly elliptic-lanceolate, chestnut brown, much compressed, 8–16 flowered; rachilla with narrow persistent wings. Glumes closely imbricate, ovate-oblong, mucronate. Nuts ellipsoid, brownish yellow, stipitate.

Fls. & Frts.: September–November. *Exsiccata*: 1228

Common, marsh loving.

Occurrence: Malpe, Mapusa, Panjim, Pudubudri.

Ecological Note: It grows in brackish water wetlands.

Cyperus halpan L. Sp. Pl. 45. 1753 (*'halpan'*); Cl. in Hook. f. Fl. Ind. Brit. India 6: 600. 1803; Aquat. Wetl. Pl. Ind. 112. 1996.

Perennial, glabrous herbs; rhizome creeping, clothed with ovate triangular scales; stem 30–120 cm long, slender, trigonous or triquetrous. Leaves usually shorter than stem, acute, erect; sheaths inflated. Umbel simple or compound; rays 4–12, very unequal, ca 8 cm long. Spikelets linear, acute, flat, 10–40 flowered; rhachilla narrow, scarcely winged. Glumes ovate-oblong, rounded on the back, obscurely mucronate, hyaline along margin. Nuts strongly angular, trigonous, smooth or slightly warty.

Fls. & Frts.: October–February. *Exsiccata*: 1243

Common, marsh loving.

Occurrence: Malpe, Pudubudri, Panjim, Ratnagiri.

Ecological Note: It grows in rice fields.

Cyperus iria L. Sp. Pl. 45. 1753; Hook. f. Fl. Brit. India 6: 606. 1893; Cook, Aquat. Wetl. Ind. 114. 1994. *C. iria* var. *paniciformis* (Franch. & Sav.) Cl. in Hook. f. *op. cit.*

Small, glabrous herbs; stem tufted, 30–50 cm long., striate. Leaves as long as stem or shorter, multiseriate, finely acuminate. Umbel decompound of many primary rays. Spikelets linear-oblong, obtuse, compressed, 6–20 flowered, yellow or pale brown. Nuts obovoid, stipitate, triquetrous, dark brown to black.

Fls. & Frts.: June–February. *Exsiccata*: 344

Common, marsh loving.

Occurrence: Malpe, Malwan, Mapusa, Pudbidri, Ratnagiri, Vaibhavwadi.

Ecological Note: It grows in rice fields.

Cyperus nutans Vahl. Enum. Pl. 2: 363. 1806; Cl. in Hook. f. Fl. Brit. India. 6: 607. 1893; Cook, Aquat. Wetl. Pl. Ind. 177. 1996.

Perennial herbs; rhizome tuberous; root fibres matted. Leaves as long as stem, linear, multiseriate. Umbel very large, decompound; primary rays reaching 25 cm long, slender, trigonous, bearing numerous secondary rays. Spikelets 6–10 flowered, red or dark brown; rachilla slender, scarcely winged, with long internodes between the glumes. Glume oblong, distant, erect, obtuse, shortly apiculate, rounded on the back, obscurely nerved, with hyaline margin. Nuts narrowly obovoid or oblong, apiculate.

Fls. & Frts.: September–November. *Exsiccata*: 456

Common, marsh loving.

Occurrence: Malwan, Malpe, Panjim, Ratnagiri.

Ecological Note: It grows in rice fields.

var. **eleusinoides** (Kunth) Haines, Bot. Bih. Or. 3: 898. 1924 [3. 941. 1961 (Repr. ed.)]; Cook, Aquat. Wetl. Pl. Ind. 117. 1996. *C. eleusinoides* Kunth, Enum. Pl. 2: 39. 1837. *C. nutans* ssp. *eleusinoides* (Kunth) T. Koyama in Gard. Bull. Singapore 30: 136. 1977.

Perennial; rootstock woody, with lateral shoots; stem 60–120 cm long, stout, triquetrous, with smooth angles. Leaves shorter than the stem, rigid, flat. Umbel erect compound, with 5–8 stout. trigonous rays, 10–20 cm long; spikes sessile or pedunculate, bracteolate. Spikelets densely imbricate, linear-oblong, reddish-brown. Nuts oblong or obovoid-oblong, apiculate.

Fls. & *Frts.*: September–February. *Exsiccata*: 6092

Common, marsh loving.

Occurrence: Malwan, Malpe, Ratnagiri, Honnawar.

Ecological Note: It grows in the brackish water wetlands.

Cyperus pilosus Vahl, Enum. Pl. 2: 354. 1806; Cl. in Hook. f. Brit. India 6: 609. 1893; Cook, Aquat. Wetl. Pl. Ind. 119. 1996; *C. pilosus* var. *obliqua* Cl. J. Linn. Soc. 21: 151. 1884 and in Hook. f. *op. cit.*, 610. *C. pilosus* var. *polyantha* Cl. *op. cit.* 148. & in Hook. f. *op. cit.*

Herbs ca 3 feet height. Leaves 6–13 mm wide, flat or faintly channelled. Bracts 10–15 cm long. Spikelets 5–10 mm long, linear. Glumes broadly ovate, mucronate. Nuts obovoid or ellipsoid, greyish, stipitate.

Fls. & Frts.: August–November. *Exsiccata*: 6066

Common, marsh loving.

Occurrence: Mapusa, Panjim, Ratnagiri, Malpe, Honnawar.

Ecological Note: It grows in the wetlands.

Cyperus rotundus L. Sp. Pl. 45. 1753; Clarke, in Hook. f. Fl. Brit. Ind. 6: 664. 1893; Cook, Aquat. Wetl. Pl. Ind. 123. 1996. NAGAR MOTHA

Perennial, glabrous, stoloniferous herbs; stem subsolitary, triquetrous at the top. Leaves narrowly linear, finely acuminate, flat, 1-nerved. Umbel simple or compound; rays 2–8, the longest reaching 8 cm long. Spikelets variable in length, linear, subacute, red-brown, 10–50 flowered; rachilla with hyaline wings. Glumes oblong, obtuse or slightly apiculate, hyaline along margins. Nuts obovate, oblong, obtusely trigonous, grey.

Fls. & Frts.: June–October. *Exsiccata*: 131

Abundant, marsh loving.

Occurrence: Kittur, Kolhapur, Ratnagiri.

Ecological Note: It grows in cultivated fields, open wetland, riverbank, margin of pools, ponds and puddles.

Eleocharis acutangula (Roxb.) Schult. in R. & S. Mant. 2: 91. 1824; Cook, Aquat. Wetl. Pl. Ind. 1: 127. 1996. *Scripus acutangulus* Roxb. Fl. Ind. 1: 126. 1893. *Eleocharis fistulosa* Schult. in R. & S. Mant. 2: 87. 1824; Cl. in Hook. f. Fl. Brit. Ind. 6: 628. 1893.

Perennial, rhizomatous, stoloniferous, tufted herb. Stem robust or slender 1–3 feet long; sheaths obliquely truncate, thinly membranous, loose. Spikelets as broad as stem or broader, greenish. Lowest glumes bractiform, broadly ovate, coriaceous, with narrow membranous margin; rachilla slender, angular, with broad concave facets between the glumes. Fertile glumes ovate-oblong or subovate, multiseriate on the back outside, often with minute brown spots. Nuts turgid unequally biconvex, ovate, constricted.

Fls. & Frts.: October–December. *Exsiccata*: 1352

Common, marsh loving.

Occurrence: Panjim, Malpe, Pudbudri.

Ecological Note: It grows in shallow water seasonally flooded pools, ponds and rice fields.

Eleocharis atropurpurea (Retz.) J. & K. Pres, Rel. Haenk. 1: 196. 1828; Cl. in Hook. f. Fl. Brit. India 6: 627. 1893; Cook, Aquat. Wetl. Pl. Ind. 127. 1996. *Scripus atropurpureus* Rezt. Obs. Bot. 5: 14. 1789.

Annual herbs; root fibrous; stem slender, densely tufted, 5–10 cm long. Leaves 0; sheaths short. Spikelets ovoid, obtuse or subacute, much broader than stem, many flowered. Glumes broadly ovate, oblong, obtuse, concave, loosely imbricate, easily detached. Nuts obovoid, biconvex, smooth, black.

Fls. & Frts.: July–January. *Exsiccata*: 390

Common, marsh loving.

Occurrence: Dhamapur lake, Karambole lake, Kagal lake, Malpe, Ratnagiri.

Ecological Note: It grows in marshes and drying margins of lakes, pools and ponds.

Eleocharis dulcis (Burm. f.) Trin. Ex Henschel, Vita Rumph. 186. 1883; Kern in Steenis, Fl. Males. 1, 7: 529. F. 36. 1974; Cook, Aquat. Wetl. Pl. Ind. 128. 1996; *Andropogon dulce* Burm. f. Fl. Ind. 219. 1768. *Eleocharis palntaginea* (Retz.) R. & S. Syst. Veg. 2: 150. 1817; Cl. in Hook. f. Fl. Brit. India 6: 625. 1893.

Robust, stoloniferous, perennial herbs; stems densely tufted, spongy. Spikelets cylindric, 2–4.5 cm long. Glumes closely imbricate, oblong; bristles 6–8, exceeding nuts. Nuts obovate, biconvex, dark brown.

Fls. & Frts.: August–March. *Exsiccata*: 1044

Common, marsh loving.

Occurrence: Malpe, Panjim, Ratnagiri.

Ecological Note: It grows on margins of pools, lakes and flooded rice fields.

Eleocharis geniculata (L.) R. & S. Syst. Veg. 2: 150. 1817; Cook, Aquat. Wetl. Pl. Ind. 129. 1996. *Scripus geniculatus* L. Sp. Pl. 48. 1753. *Eleocharis capitata* R. Br. Prodr. 225. 1810 *non Scripus capitatus* L. 1753; Cl. in Hook. f. 6: 627. 1893.

Erect herbs, densely tufted. ca 15 cm high. Spikelets 3–7 × 3–7 mm, globosely ovoid, greenish-yellow or brown. Nuts 1 mm long, globose-ovoid, brown, smooth, shining.

Fls. & Frts.: September–January. *Exsiccata*: 1045

Common, marsh loving.

Occurrence: Malwan, Ratnagiri,Vaibhavwadi,

Ecological Note: It grows in rice fields.

Eleocharis retroflexa (Poir.) Urb. Symb. Antill. 2: 165. 1900; Cook, Aquat. Wetl. Pl. Ind. 131. 1996. *Scirpus retroflexus* Poir. in Lam. Encycl. 6: 753. 1804. *Elaeocharis chaetaria* R. & S. Syst. Veg. 2: 154. 1817; Cl. in Hook. f. Fl. Brit. Ind. 6: 629. 1893.

Annual, tufted, gregarious herbs; stem angular often curved. Sheaths reddish at the base. Spikelets ovoid, obtuse, usually tinged with red, terete or compressed. Lowest glumes boat-shaped, empty, ovate, membranous, 3-nerved. Bristles 3, unequal, white to yellowish, scabrous, longer than nuts. Nut obovoid, conspicuously cancellate.

Fls. & Frts.: January–March. *Exsiccata*: 411

Common, marsh loving.

Occurrence: Malwan, Malpe.

Ecological Note: It grows on marshy margin of shallow water ponds and ditches and in flooded rice field.

Fimbristylis aestivalis (Retz.) Vahl, Enum. Pl. 2: 288. 1806; Cl. in Hook. f. Fl. Brit. Ind. 6: 637. 1893; Cook, Aquat. Wetl. Pl. Ind. 135. 1996. *Scirpus aestivalis* Retz. Obs. Bot. 4: 12. 1786.

Tufted herbs; stem angular, 5–15 cm long, erect, trigonous, filiform, striate. Leaves shorter than stems, narrow, acute. Umbels compound or subdecompound, lax, with many slender erect rays; bracts 3–5, narrow, linear, acute, brown; two glumes empty. Spikelet few, oblong. Glumes oblong, puberulous, suberect, back strongly keeled. Nuts broadly elliptic, compressed.

Fls. & Frts.: September–June. *Exsiccata*: 1055

Common, marsh loving.

Occurrence: Malpe.

Ecological Note: It grows in marshy places.

Fimbristylis argentea (Rottb.) Vahl, Enum. Pl. 2: 294. 1806; Cl. in Hook. f. Fl. Brit. Ind. 6: 640. 1893; Cook, Aquat. Wetl. Pl. Ind. 137. 1996.

Annual densely tufted herbs; root fibrous; stems many, 10–25 cm long, filiform, trogonous, striate. Leaves shorter than stem, slender, glabrous, flexuous, canaliculated; sheaths short, glabrous, red dotted. Spikelet cylindrical in heads, sessile. Glumes closely imbricate, ovate, obtuse, not cuspidate, 3-nerved. Nuts orbicularly ovoid, compressed, stipitate.

Fls. & Frts.: February–March.

Common, marsh loving.

Occurrence: Malpe, Panjim, Pudbudri.

Ecological Note: It grows in swampy areas.

Fimbristylis bis-umbellata (Forssk.) Bub. Dodec. 30. 1850; Cook, Aquat. Wetl. Pl. Ind. 137. 1996. *F. bis-umbellatus* Forssk. Fl. Aegypt. Arab. 15. 1755. *F. dichotoma* sensu Cl. in Hook. f. Fl. Brit. Ind. 6: 635. 1893.

Annual herbs; root fibrous; stem 5–15 cm long, tufted, filiform, striate, smooth. Leaves shorter than stem, narrow, tapering to a fine point; sheaths softly villous. Umbel laxly compound or decompound; rays few to many; bracts filiform. Spikelets ellipsoid, many flowered; rachilla angled with large pits. Glumes ovate-oblong, reddish-brown. Nuts obovate, pale to dark brown.

Fls. & Frts.: May–December. *Exsiccata*: 41

Common, marsh loving.

Occurrence: Mapusa, Panjim, Udupi.

Ecological Note: It grows on margins of ponds and puddles.

Fimbristylis complanata (Retz.) Link, Hort. Bot. Berol. 1: 292. 1827; Cl. in Hook. f. Fl. Brit. Ind. 6: 646. 1893; Cook, Aquat. Wetl. Pl. Ind. 138. 1996. *Scripus complanatus* Retz. Obs. Bot. 5: 14. 1996.

Perennial hebs; rhizomes short; stem 20–55 cm long, deeply grooved, ribbed. Leaves crowded at the base, flat, suddenly narrowed at the tip, obtuse or subacute, scabrid on margins. Umbel decompound, effuse, with few or many flattened rays loosely corymbosely branched. Spikelets ovoid-oblong, few flowered, with more or less distichously arranged glumes. Nuts stipitate, obtusely trigonous, minutely warted.

Fls. & Frts.: July–December. *Exsiccata*: 42

Common, marsh loving.

Occurrence: Malpe, Panjim.

Ecological Note: It grows on riverbank and swampy grassland.

Fimbristylis dipsacea (Rottb.) Cl. in Hook. f. Fl. Brit. Ind. 6: 635. 1893; Cook, Aquat, Wetl. Pl. Ind. 141. 1996. *Scirpus dipsaceus* Rottb. Decr. & Ic. Pl. 56, t. 12. F. 1. 1773. *Echinolytrum dispaceum* (Rottb.) Desv. in J. Bot. 1: 21, t. 1. 1808.

Annual herbs, densely tufted, 2–15 cm high. Leaves reduced to short blade; sheath 2–4 per culm, blades narrowly linear to setaceous. Corymbs simple, umbeliform or a cluster of 2–3 spikelets or reduced to a single spikelet. Achenes linear-oblong, thickly biconvex.

Fls. & Frts.: May–June.

Common, marsh loving.

Occurrence: Shirur river.

Ecological Note: It grows on bank of river and stream.

Fimbristylis ferruginea (L.) Vahl, Enum. Pl. 2: 291. 1806; Cl. in Hook. f. Fl. Brit. Ind. 6: 638. 1893; Cook, Aquat. Welt. Pl. Ind. 142. 1996. *Scirpus ferrugineus* L. Sp. Pl. 50. 1753.

Perennial herbs; root fibres stout and slender; stem ca 60 cm high, subtrigonus. Leaves absent or few, short, narrowly linear;sheath pubescent, thin. Umbel simple or subcompound. Spikelets ellipsoid, pale reddish brown, lowest bracts usually bractiform. Glumes erect or slightly curved, broadly ovate, subacute, concave. Nuts biconvex, shortly stipitate, umbonate.

Fls. & Frts.: September–February. *Exsiccata*: 1462

Common, marsh loving

Occurrence: Dhamapur, Malpe.

Ecological Note: It grows in muddy places.

Fimbristylis littoralis Gaud. in Freyc. Voy. Bot. 413. 1826; Cook, Aquat. Wetl. Pl. Ind. 143. 1996. *F. miliacea* sensu Benth Fl Austral. 316. 1878; Cl. in Hook. f. Fl. Brit. Ind. 6: 644. 1893 p.p. omne non Vahl, 1806 non *Scirpus miliaceus* L.

Annual, glabrous, leafy, herbs; stem tufted, 15 cm high. Leaves distichous, linear, finely acuminate, flaccid, with scaberulous margin. Umbel decompound; rays unequal, suberect or spreading, filiform; bracts filiform, shorter than umbel. compressed. Spikelets subglobose, subcylindric, obtuse, brown. Glumes closely imbricate, ovate, obtuse, cymbiform, membranous. Nuts obovoid, white, becoming brown, verrulose.

Fls. & Frts.: June–August. *Exsiccata*: 6067

Common, marsh loving.

Occurrence: Malpe, Panjim, Ratnagiri.

Ecological Note: It grows on wet margin of lakes and pools and in swampy places.

Fimbristylis miliacea (L.) Vahl, Enum. Pl. 2: 287. 1806 emend. S. T. Blake in J. Arnold Arbor. 35: 216. 1954; Cl. in Fl. Brit. Ind. 6: 644. 1893; Cook, Aquat. Wetl. Pl. Ind. 145. 1996. *Scripus miliaceus* L. Syst. Veg. ed. 10, 2: 229. 1759; Clarke, in Hook. f. *op. cit.*

Annual herbs; roots fibrous; stems tufted, 30–90 cm high. Leaves linear, tapering to a firm point. Umbel decompound; rays unequal, suberect or spreading, filiform. Spikelets ellipsoid or ovoid, obtuse, greenish-brown. Nuts obovoid, shortly stipitate, obtusely trigonus, slightly umbonate.

Fls. & Frts.: June–February. *Exsiccata*: 415

Common, marsh loving.

Occurrence: Malpe, Panjim, Ratnagiri.

Ecological Note: It grows in wet places, margin of drying ponds and in rice fields.

Fimbristylis ovata (Burm. f.) Kern in Blumea 15: 126. 1967; Cook, Aquat. Wetl. Pl. Ind. 145. 1996. *Carex ovata* Burm. f. Fl. Ind. 194. 1768. *Fimbristylis monostachyos* (L.) Hassk. Pl. Jav. Rar. 61. 649. 1848 (*monostachya*); Cl. in Hook. f. Fl. Ind. 6: 649. 1893.

Densely tufted, leafy herbs; stems 15–30 cm long, very slender, filiform, striate. Leaves crowded on the swollen base of the stem, filiform, acute or acuminate. Spikelets solitary on the stem, compressed, ovate, acute, shining, pale straw coloured; lower two glumes empty, longer than other, cuspidate. Nuts pyriform, sessile, umbonate, tubercled.

Fls. & Frts.: August–November. *Exsiccata*: 6093

Common, marsh loving.

Occurrence: Malpe, Mapusa, Panjim, Ratnagiri

Ecological Note: It grows in wetlands and often found in marshes.

Fimbristylis polytrichoides (Retz.) Vahl, Enum. Pl. 2: 248. 1806; Cl. in Hook. f. Fl. Brit. Ind. 6: 632. 1893; Cook, Aquat. Wetl. Pl. Ind. 145. 1996. *Scirpus polytrichoides* Rezt. Obs. Bot. 4: 11. 1786.

Perennial glabrous herbs; stems 5–25 cm long, filiform, leafy at the base, trigonous, striate. Leaves as long as or longer than stem, slender, wiry, acute, terete, or trigonous. Spikelets solitary, oblong, ellipsoid, many flowered; glumes linear-oblong or cymbiform, acute, minutely apiculate, straw coloured or white. Nut turbinate, sessile, much compressed, yellow, dotted with white spot.

Fls. & Frts.: August-October. *Exsiccata*: 486

Common, marsh loving.

Occurrence: Malpe, Roha, Panjim, Ratnagiri.

Ecological Note: It grows in marshes and in wet areas.

Fimbristylis schoenoides (Retz.) Vahl, Enum. Pl. 2: 286. 1806; Cl. in Hook. f. Fl. Brit. Ind. 6: 634. 1893; Cook, Aquat. Wetl. Pl. Ind. 147. 1996. *Scirpus schoenoides* Retz. Obs. Bot. 5: 14. 1789.

Glabrous herbs; stem 10–45 cm long, tufted, slender, obtusely angled, striate. Leaves shorter than stem, narrowly linear, acute, striate above, dotted beneath, with recurved spinolose margin. Spikelets solitary or 2–5 in irregular umbel, sessile or on slender pedicels. Glumes deciduous, loosely imbricate, orbicular, obtuse, shortly apiculate, membranous. Nuts turbinate, compressed, biconvex, rounded at apex, smooth, shortly stipitate.

Fls. & Frts.: August–October. *Exsiccata*: 6091

Common, marsh loving.

Occurrence: Malpe, Panjim, Ratnagiri.

Ecological Note: It grows in wet grasslands, seasonally inundated areas and margin of ponds, puddles and pools.

Fimbristylis tetragona R. Br. Prodr. 226. 1810; Cl. in Hook. f. Fl. Brit. Ind. 6: 631. 1893; Cook, Aquat. Wetl. Fl. Ind. 149. 1996.

Perennial, glabrous herbs; stem 15–60 cm long, tufted, slender, quadrangular, ribbed. Leaves absent; sheaths usually brown. Spikelets solitary, ovoid-conic, terete, obtuse; rachilla thickly studded with deep tetragonal pits with raised edges. Lowest glumes deciduous. Nuts linear, oblong or subcylindric, narrow, tapering towards the base, stipitate, tuberculate.

Fls. & Frts.: September–October. *Exsiccata*: 1349

Common, marsh loving.

Occurrence: Malpe, Ratnagiri, Wannawar, Ramthirth.

Ecological Note: It often grows in shallow water ponds and puddles and in marshy places.

Fuirena ciliaris (L.) Roxb. Fl. Ind. 1: 184. 1820; Cook, Aquat. Wetl. Pl. Ind. 153. 1996. *Scirpus ciliaris* L. Mant. Alt. 182. 1771. *Fuirena glomerata* Lam. Illustr. 1: 150. 1791; Cl. in Hook. f. Fl. Brit. Ind. 6: 666. 1893.

Annual herbs; stem 10–15 cm long, slender, erect, leafy. Leaves linear-lanceolate, acuminate, 3–5-nerved, hairy; sheaths long, closed, straite. Spikelets clustered, ovoid, oblong; lowest glumes longest, empty. Nuts ellipsoid or obovoid, stipitate, trigonous with acute angles, beaked.

Fls. & Frts.: September–February. *Exsiccata*: 5952

Common, marsh loving.

Occurrence: Malpe, Malwan, Panjim, Ratnagiri, Udupi.

Ecological Note: It grows in open wetlands and swampy places.

Kyllinga brevifolia Rottb. Descr. & Ic. 13, t. 4, f. 3. 1753; Cl. in Hook. f. Fl. Brit. Ind. 6: 588. 1893; Cook, Aquat. Wetl. f. Fl. Ind. 156. 1996. *Cyperus brevifolius* (Rottb.) Hassk. Cat. Hort. Borg. 24. 1844.

Perennial herbs, ca 30 cm high. Leaves 13–17 cm, sub-basal shorter or as long as stem. Inflorescence head, solitary, terminal. Spikelets 2–5 × 1 mm, ovoid-lanceolate, compressed, greenish-brown. Nuts globosely ovoid, apiculate, vertically verrucose.

Fls. & Frts.: June–February. *Exsiccata*: 1999

Common, marsh loving.

Occurrence: Malwan, Malpe, Panjim, Ratnagiri.

Ecological Note: It grows on marshy margins of ponds, puddles and pools.

Mariscus clarkei (T. Cooke) T. Koyama in J. Jap. Bot. 51: 313. 1976 & Gard. Bull. Singapore 30: 158. 1977; Cook, Aquat. Wetl. Pl. Ind. 163. 1996. *Cyperus clarkei* T. Cooke, Fl. Pres. Bombay 2: 873. 1908 [3: 387. 1958 (Repr. ed.)]. *Mariscus bulbosus* Cl. in Hook. f. Fl. Brit. Ind. 6: 620. 1893.

Small sedge. Rootstocks tuberous, clothed with oblong-lanceolate, striate, pale yellow scales. Leaves linear, acute; lower sheaths much inflated, striate brown. Umbel simple, with very short rays or subcapitate. Spikelets densely packed, spreading nearly at right angles, bearing 1-nut. Nut bearing glumes ovate, acute, multiseriate, straw coloured. Nuts ellipsoid, trigonous, brown.

Fls. & Frts.: August–October. *Exsiccata*: 6032

Common, marsh loving.

Occurrence: Malpe, Malwan, Panjim, Pudbudri.

Ecological Note: It grows in seasonally flooded areas.

Mariscus compactus (Retz.) Boldingh, Zakfl. Landb. Java 77. 1916; Cook, Aquat. Wetl. Pl. Ind. 163. 1996. *Cyperus compactus* Retz. Obs. Bot. 5: 10. 1789. *C. dilutus* Vahl, Enum. Pl. 2: 357. 1806. *Mariscus microcephalus* Presl. Rel. Haenk. 1: 182. 1830.

Rhizomatous, perennial herbs; stem terete or obtusely trigonous. Leaves equal to or longer than stems, rigid, strongly channelled, glaucous. Umbel compound or decompound; rays spreading; bracts reflexed; spike globose; spikelets linear-lanceolate, subterete; glumes convolute, oblong-lanceolate, keel 3-nerved, green. Nuts oblong-linear, trigonous.

Fls. & Frts.: October–November. *Exsiccata*: 6044

Common, marsh loving.

Occurrence: Malpe, Panjim, Ratnagiri.

Ecological Note: It grows in seasonally flooded areas, wet rice fields, marshes and ditches.

Mariscus squarrosus (L.) Cl. in Hook. f. Fl. Brit. India 6: 623. 1893; Cook, Aquat. Wetl. Pl. Ind. 166. 1996. *Cyperus squarossus* L. in Torner, Cent. Pl. 2: 6. 1756. *Cyperus aristatus* Rottb. Descr. Pl. Rar. Progr. 22. 1772; Cl. in Hook. f. *op. cit* 606. *Mariscus aristatus* (Rottb.) Tang & Wang in Fl. Rep. Pop. Sin. 11: 178, t. 59, ff. 12–13. 1961.

Herbs, ca 10 cm high. Leaves 5–8 cm long, narrowly linear. Inflorescence umbellate, umbel 2–5; rays often capitate. Spikelets oblong-linear, greenish-brown; glumes aristate. Nuts 1 × 1 mm, obovoid, trigonous, stipitate.

Fls. & Frts.: June–November. *Exsiccata*: 1355

Common, marsh loving.

Occurrence: Madgoan, Malpe, Malwan, Vaibhawadi, Panjim.

Ecological Note: It grows on edges of tanks, pools, streams, ditches and in rice fields.

Pycreus flavidus (Retz.) T. Koyama in J. Jap. Bot. 51: 316. 1976; Cook, Aquat. Wetl. Pl. Ind. 168. 1996. *Cyperus flavidus* Retz. Obs. Bot. 5: 13. 1789. *Pycreus globosus* (All.) Reichb. Dt. 692. 1976. *P. capillaris* (Koen. ex Roxb.) Nees ex Cl. in Hook. f. Fl. Brit. Ind. 6: 591. 1893.

Herbs, annual, erect. Leaves shorter than heads, slender. Inflorescence of simple umbels of contracted heads. Spikelets 6 × 3 mm, red or brown. Nuts minutely tuberculate.

Fls. & Frts.: July–March. *Exsiccata*: 383

Common, marsh loving.

Occurrence: Malpe, Panjim, Pudubudri.

Ecological Note: Grows in wet, marshy and swampy places.

Pycreus pumilus (L.) Nees ex Cl. in Hook. f. Fl. Brit. Ind. 6: 591. 1893; Cook, Aquat. Wetl. Pl. Ind. 170. 1996. *Cyperus pygmaeus* Retz. Obs. Bot. 4: 9. 1786; Cl. in Hook. f. 6: 619. 1893. *C. pusillus* Vahl, Enum. Pl. 2: 303. 1806; Cl. in Hook. f. op. cit. *Pycreus nitens* (Retz.) Nees in Nova Acta Nat. Cur. 19, Suppl. 1: 53. 1843; Cl. in Hook. f. op. cit. 519.

Annul herbs; stem tufted, 3–36 cm long, slender. Leaves linear, shorter than stem, 1-nerved, acute. Inflorescence a compact umbel of short spikes. Spikelets 20–50 flowered, much compressed; rachilla stout, not winged. Glumes ovate, boat shaped, 2-fid, with narrow herbaceous keel. Nuts ellipsoid-oblong or sub-ovoid, brown.

Fls. & Frts.: August–February. *Exsiccata*: 497

Common, marsh loving.

Occurrence: Malpe, Panjim, Pububudri, Ratnagiri.

Ecological Note: It grows in wet, marshy and swampy places.

Rhynchospora corymbosa (L.) Britt. In Trans. New York Acad. Sci. 11: 84. 1892; Cook, Aquat. Wetl. Pl. Ind. 173. 1893. *Scirpus corymbosus* L. in Torner, Cent. Pl. 2: 7. 1756. *Rhynchospora aurea* Vahl, Enum. 2: 229. 1806; Cl. in Hook. f. Fl. Brit. India 6: 671. 1893.

Rhizomatous, perennial herbs; stems leafy throughout, triquerous with smooth angles. Leaves flat, with prominent midrib. Inflorescence paniculate, corymbiform. Spikelets lanceolate, brown. Nuts sessile, obovate-oblong, brown, beaked; beak conical, grooved on both sides.

Fls. & Frts.: November–December. *Exsiccata*: 6033

Common. marsh loving.

Occurrence: Lolium, Ratnagiri, Vaibhavwadi.

Ecological Note: It grows in open wetlands among grasses.

Schoenoplectus juncoides (Roxb.) Palla in Bot. Jahrb. Syst. 10: 299. 1888; Cook, Aquat. Wetl. Pl. Ind.177. 1893. *Scirpus juncoides* Roxb. Fl. Ind. 1: 218. 1820. *S. erectus sensu* Cl. in Hook. Fl. Brit. Ind. 6: 656. 1893.

Annual herbs; stem slender, terete, tufted. Leaves reduced to bladeless sheaths. Inflorescence pseudolateral, capitate; bracts much shorter than stems. Spikelets sessile, ovoid-elliptic. Glumes broadly ovate, concave, reddish-brown, keeled at apex, obtuse; dorsiventrally compressed, biconvex, broadly ovate, black.

Fls. & Frts.: February–March. *Exsiccata*: 5323

Common, marsh loving.

Occurrence: Malpe, Pudbudri.

Ecological Note: It grows in seasonally flooded areas and water logged paddy fields.

HYDATELLACEAE

Trithuria konkanensis Yadav et Janarthanam in Rheedea 4:18, f. 1–6. 1994; Aqua Planta 3: 91–97. 1995; Cook, Aqua. Wetl. Pl. Ind. 215, f. 222. 1996.

Annual, acaulescent, tufted herbs, 1–1.5 cm, green to red in colour. Inflorescence capitulum, stalked or sessile, peduncle upto 2 mm long. Flowers unisexual, naked. Male flower solitary, represented by single stamen at the center of capitula surrounded by female flowers; filament stout, upto 7 mm long, dark pink; anther basifixed, oblong-elliptic, tertrasporangiate. Female flowers 15–20 per capitulum, each represented by solitary pistil with 4–6 stigmatic hairs; ovary ovoid, triquetrous, stalked; stigmatic hairs 2–6, uniseriate, upto 5 mm long. Fruit *ca* 0.4 × 0.27 mm, ovoid, triquetrous, indehiscent. Seeds translucent, brown with dark tip.

Fls. & Frts.: July–September.

Common, semiaquatic hydrophyte.

Occurrence: Achirne, Lolium, Manipal, Ratnagiri. Endemic.

Ecological Note: It grows on lateritic plateaus in monsoon season.

JUNCACEAE

Juncus prismatocarpus R. Br. Prodr. Fl. Nov. Holland. 259. 1890; Hook. f. Fl. Brit. Ind. 6: 395. 1892; Cook, Aquat. Wetl. Pl. Ind. 223. 1996.

Tufted herbs. Leaves flat, terete, transversely septate, reduced; sheaths scarious. Inflorescence with compound cymes of head, bracteate. Perianth lobes 6, linear, persistent. Fruit capsule.

Fls. & Frts.: March–April. *Exsiccata*: 6091

Occasional, marsh loving.

Occurrence: Honnawar

Ecological Note: It grows in marshy places near water bodies and wet paddy fields.

POACEAE

Coix aquatica Roxb. Fl. Ind. 3: 571. 1832; Hook. f. Fl. Brit. India 7: 100. 1896; Bor, Grass. Ind. 264. 1960; Cook, Aqua. Wetl. Pl. India 288. 1996.

Annuals or perennials. Leaves linear lanceolate, ensiform, long acuminate; ligule membranous. Spike terminal; peduncle solitary or in pair, drooping. Male spikelets ovate-lanceolate or broadly ovate, arranged in 6 longitudinal rows; female spikelets solitary at base of male spike.

Fls. & Frts.: October–February. *Exsiccata*: 5941

Common, Rooted emergent.

Occurrence: Throughout coastal plains.

Ecological Note: It grows in fresh water bodies forming large floating mats.

Leersia hexandra Swartz, Prodr. Veg. Ind. Occ. 21. 1788; Hook. f. Fl. Brit. India 7: 94. 1896; Cook, Aqua. Wetl. Pl. India 302. 1996.

Ligules membranous, 4–9 mm long; leaf blades linear-lanceolate. Inflorescence lax panicle, with flexuous branches. Spikelets 1-flowered, strongly flattened laterally, 3–4.5 mm long, yellowish green, awnless, sessile or nearly so. Glumes reduced to entire or 2-lobed thickened rim. Lemmas all fertile. Paleas linear-oblong, 2–3 mm long.

Fls. & Frts.: Throughout year. *Exsiccata*: 5944

Occasional, rooted emergent.

Occurrence: Kolhapur, Sawantwadi

Ecological Note: It grows in rice fields and in shallow water tanks

Oryza rufipogon Griffith Notul. 3: 5. 1851; Bor, Grass. Ind. 605. 19?0; Cook, Aqua. Wetl. Pl. India 304. 1996.

Spikelets disarticulating above the glumes, 6–11 mm long, deciduous. Glumes 0.3 mm long. Sterile lemmas triangular to linear-lanceolate. Fertile lemma obovate to lanceolate, 1.25–7.5 mm long. Anthers 4–6.2 mm long, yellow or brown. Caryopses oblong, cylindrical, reddish-brown.

Fls. & Frts.: July–August. *Exsiccata*: 5317

Common, rooted emergent.

Occurrence: Lolium, Manipal.

Ecological Note: It grows in temporary ponds and puddles on lateritic plateaus.

Paspalidium geminatum (Forssk.) Stapf in Prain, Fl. Trop. Afr. 9: 583. 1920; Bor, Grass. Ind. 333. 1960; Subra. Aqua. Ang. 172, f. 60. 1962; Cook, Aqua. Wetl. Pl. India 307. 1996. *Panicum geminatum* Forssk. Fl. Aegypt.-Arab. 18. 1775. *P. paspaloides* Pers. Syn. 1: 81. 1805; Hook. Fl. Brit. India &: 30. 1896.

Rhizome creeping, spongy when in water, ca 1 cm diameter. Culms prostrate, rooting below, 10–50 cm long. Ligules 1mm long; Leaf blades acute to acuminate. Racemes 10–30, ovoid. Spikelets 10–30, ovoid, 1.6–2.6 mm long. Lower glume truncate; upper glume 7–nerved. Lower floret with a palea, its lemma as long as the spikelet, with nerves not raised; lemma of upper floret granulose. Caryopses ellipsoid.

Fls. & Frts.: September–March. *Exsiccata*: 5943

Common, rooted emergent.

Occurrence: Throughout coastal Plains.

Ecological Note: It grows in permanent wet places such as margin of lakes, ponds and pools.

TYPHACEAE

Typha angustifolia L. Sp. Pl. 971. 1753; Backer in Steenis Fl. Males. 1, 4: 243, f. 1. 1951. *Typha angustata* Bory & Chaub. Exped. Sci Moree Bot. 2, 1. 338. 1822; Hook. f. Fl. Brit. India 6: 489. 1893; Subr. Aquat. Ang. 74. 1962. PANKANIS.

Robust, tufted, perennial herbs. Leaves linear, thick, sub-cylindric above the sheath. Male and female flowers separated by interval ca 30 cm. Male flowers slender, paler than female ones. Nutlets small, ellipsoid, membranous, indehiscent, brown.

Fls. & Frts.: March–October. *Exsiccata*: 5947

Abundant, Marsh loving.

Occurrence: Udupi, Kittur, Ratnagiri.

Ecological Note: Along the margin of fresh water bodies such as tanks, pools, ponds etc.

XYRIDACEAE

Xyris indica L. Sp. Pl. 62. 1753; Hook. f. Fl. Brit. India. 6: 304. 1892; Steenis, Fl. Males. 1. 4: 373, f. 1N, 4–5. 1953; Cook, Aquat. Wetl. Pl. Ind. 368. 1994. DADUMARI

Scapigerous annual herb. Leaves usually shorter than or sometimes as long as scape, spongy, linear, narrowed to an obtuse or acute tip. Scape stout, grooved, acutely angled. Head ovoid-subglobose or ellipsoid. Flowers bright yellow in ellipsoid spikes; bracts many, orbicular or cuneately obovate, usually broader than long, dark red-brown, coriaceous. Capsules ovoid. Seeds minute, ellipsoid, very strongly ribbed.

Fls. & Frts.: October–January. *Exsiccata*: 1015

Common, marsh loving.

Occurrence: Malpe, Pavas, Sawantwadi, Vaibhavwadi.

Ecological Note: It grows in open wetlands such as plateaus among the grasses.

Riverine Plants

LECYTHIDACEAE

Barringtonia acutangula Gaertn. Fruct. 2: 97, t. 101. 1791; Cl. in Hook. f. Fl. Brit. India 2: 508. 1879.

Glabrous trees; branchlets slender. Leaves obovate-oblong or elliptic, 7.5–15 × 2.2–6.5 cm, narrowed at base, rounded or subacute at apex, petioles 3–6 mm long. Flowers fragrant, 6–12 mm across, in 15–30 cm long racemes, pedicels 1.5–3 mm long; bracteoles linear-lanceolate as long as the teeth; teeth 4 oblong rounded or ciliolate. Petals red, ovate-oblong, 5 mm long. Filaments much exerted. Fruits bluntly quadrangular, 2–3 × 1–2 cm, truncate at both ends; crowned by persistent calyx. Seeds ovoid, black.

Fls. & Frts.: Aug.–Sept. *Exsiccata*: SUK 2303

Frequent, riparine plant.

Occurrence: Ratnagiri, Maxem, Panjim, Bincholim, Manglore, Sawantwadi.

Ecological Note: It grows along the streams.

BORAGINACEAE

Rotula aquatica Lour. Fl. Cochinch.121. 1790. *Rhabdia lycicoides* Mart. Nov. Gen. & Sp. Pl. 2: 137, t. 195. 1827; Cl. in Hook. f. Fl. Brit. India 4: 145. 1883.

Small shrubs. Leaves sessile, spathulate, base cuneate, apex rounded or shortly apiculate, hairy. Flowers axillary, shortly pedicelled; calyx hairy, deeply divided; corolla lobes oblong, spreading. Drupes subglobose, tipped with long, persistent style.

Fls. & Frts.: Sept.–Jan. *Exsiccata*: SUK 2304

Frequent, rivarine Plant.

Occurrence: Chiplun, Pophali.

COMBRETACEAE

Terminalia arjuna (Roxb. *ex* DC.) Wt. & Arn., Prodr. 314. 1834; Cl. in Hook. f. Fl. Brit. India 2: 447. 1878. *Pentaptera arjuna* Roxb. Fl. Ind. 2: 438. 1832. ARJUN SADADA

A large tree, with a very thick trunk and horizontally spreading branches, bark thick, smooth, greenish white, flaking off into large flat pieces. Leaves usually subopposite, 10–15 × 5–7 cm, oblong or elliptic-oblong, obtuse or subacute pale green above, pale brown beneath, shallowly crenate serrate in the upper part or sometimes throughout, base rounded or cordate often unequal sided; petioles 0.6–0.9 cm long with one or usually 2–3 prominent glands. Flowers sessile in short axillary spikes or in terminal panicles; bracteoles linear lanceolate, shorter than flower, cauducous. Calyx glabrous, 4 mm long, mouth broadly campanulate teeth triangular Ovary quite glabrous. Disk clothed with yellowish or reddish hairs. Stamens much exerted. Drupes 2.5–5 cm long, ovoid or obovoid-oblong, fibrous woody, glabrous dark brown with 5 hard projecting wings striated with numerous curved veins.

Fls. & Frts.: Apr.–Nov.

Frequent, riparine plant.

Occurrence: Ratnagiri, Maxem, Panjim, Bincholim, Manglore, Sawantwadi.

Ecological Note: It grows on stream banks and riverbeds.

EUPHORBIACEAE

Homonoia riparia Lour. Fl. Cochinch. 2: 637. 1790; Hook. f. Fl. Brit. India 5: 455. 1887. *Adelia neriifolia* Heyne ex Roth, Nov. Pl., Sp. 375. 1821. MACHIM

A rigid, evergreen shrub; branchlets, young leaves and inflorescence pubescent. Leaves numerous, all closely set, 7–15 × 1–2 cm, linear-oblong or linear-lanceolate, acute, apiculate, entire or more or less toothed, glabrous above. Male flowers: Sepals 3, broadly elliptic, acute, concave, glabrous, 0.3 cm long. Stamens numerous, in globose heads; anthers red. Female flowers: Sepals usually 5, narrowly ovate, acuminate, 2–3 mm long. Ovary pubescent; styles 3, shortly connate at base. Capsules globose, 0.3 cm across, tomentose. Seeds rounded on back, slightly angular on inner face, smooth, yellowish-brown.

Fls. & Frts.: Jan.–Jun. *Exsiccata*: SUK 2203

Rare, riparine plant.

Occurrence: Ratnagiri, Maxem, Panjim, Manglore, Sawantwadi.

Ecological Note: It grows in rocky riverbeds.

Phyllanthus lawii Grah., Cat. Bombay Pl. 181. 1839; Hook. f. Fl. Brit. India 5: 290. 1887.

Medium sized shrubs; branches drooping, strongly ribbed, often prickly, dark purplish brown. Leaves distichous on short leafy branches overlapping, linear-oblong, 3–6 × 2–4 cm, rounded or subcordate at base, entire, rounded and apiculate at apex, glabrous; petioles scarsely 0.5 mm long; stipules lanceolate, subulate, 1 mm long. Flowers minute, axillary, the males 1–3 together, females solitary. Parianth segments 6, linear-oblong, 2–2.5 mm long, obtuse. Stamens 3; filaments united into column. Style distinct, bifid at apex. Fruits depressed globose, 2–2.5 mm in diameter, brown, smooth. Seeds trigonous, 2 mm long, rounded on back, glabrous.

Fls. & Frts.: Mar.–Apr. *Exsiccata*: SUK 2229

Common, riparine plant.

Occurrence: Tillari.

Ecological Note: It grows in the riverbeds.

MELASTOMACEAE

Melastoma malabathricum L., Sp. Pl. 390. 1753; Cl. in Hook. f. Fl. Brit. India 2: 523. 1879.

A much branched shrub; young branches, petioles and nerves of leaves beneath densely clothed with appressed flat lanceolate, acuminate. Leaves somewhat rigid, 5–11 × 1.5–4 cm, ovate-oblong or lanceolate-oblong, acute strigose above with appressed thick hairs, softly hairy betwen nerves. Flowers in terminal corymose, 1–5 flowered panicles, pedicels about half as long as the calyx tube; tube broadly campanulate; lobes lanceolate acute, deciduous intermediate teeth subulate, bistle pointed. Petals 2–2.5 cm long, obovate rounded or truncate at apex narrowed at base, filaments yellow. Anthers alternately yellow and red. Fruit 1–2 cm long, broadly obovoid, truncate.

Fls. & Frts.: Oct.–Mar. *Exsiccata*: SUK 2379

Frequent, riparine plant.

Occurrence: Ratnagiri, Maxem, Panjim, Manglore, Sawantwadi.

Ecological Note: It grows on stream banks and riverbeds.

MYRTACEAE

Syzigium cumini (L.) Skeels, U.S. Dept. of Agriculture Bur. Pl. Ind. Bull. 2: 248. 1912. *Myrtus cumini* L., Sp. Pl. 471. 1753. *Eugenia jambolana* Lamk., Encycl. 3: 198. 1789; Duthie in Hook. f. Fl. Brit. India 2: 499. 1879. JAMBHAL

Large trees; bark light coloured thick rough, exfoliatinmg branches terete. Leaves coriaceous variable, 6–11 × 4–6 cm, lanceolate elliptic-oblong or broadly ovate-elliptic, acute subobtuse or shortly narrowed at base. Flowers dirty white fragrant about 2–3 mm across sessile or shortly pedicelled crowded in heads on the ends of laxly panicled cymes rising from the branches below the leaves. Fruit dark violet globose or ellipsoid, smooth, variable in size.

Fls. & Frts.: March–July. *Exsiccata*: SUK 1915

Frequent, riparine plant.

Occurrence: Honnavar, Maxem, Panjim, Manglore, Sawantwadi, Udupi.

Ecological Note: It grows on stream banks and riverbeds.

Syzigium heyneanum (Duthie) Wall. *ex* Gamble, Fl. Madras 1: 482. 1919. and 1; 341. 1957 (Repr. ed.) *Eugenia heyneana* Duthie in Hook. f. Fl. Brit. India 2: 500. 1879. *Syzigium salicifolium* Grah. Cat. 73. 1839.

Large shrub or small tree. Leaves 7–10 × 2–2.5 cm, oblong lanceolate, subobtuse, glabrous and shining pellucid dotted, base tapering, main nerves numerous uniting in an intramarginal vein, petiole 3–9 mm long. Flowers 0.5–0.6 cm across, whitish, sessile or shortly pedicellate in small heads at the ends of branches of cymes which are lateral from the scars of fallen leaves, rarely axillary. Calyx 3–9 mm long, externally rugulose, funnel-shaped. Petals calyptrate. Fruits oblong-ellipsoid, 1–2 cm long, crowned with the cup shaped calyx limb.

Fls. & Frts.: Mar.–Jul. *Exsiccata*: SUK 1787

Frequent, riparine plant.

Occurrence: Ratnagiri, Maxem, Manipal, Sawantwadi, Roha.

Ecological Note: It grows on stream banks and riverbeds.

SALICACEAE

Salix tetrasperma Roxb. Pl. Corom. 1: 66. t. 1795; Hook. f. Fl. Brit. India.5: 626. 1888. WALUNJ

Much branched trees; young shoots silky, glabrescent. Leaves alternate, lanceolate, 5–15 × 1.5–3cm, drooping, acute or rounded at base, entire or serrulate, acute or acuminate. Male flowers sessile, in axillary, 5–10cm long catkin; bracts ovate, concave, silky villous on the back. Disk of 2 yellow glands, stamens 5–10. Female flower pedicellate, in the axils of oblong or obovate-oblong, hairy bracts. Ovary glabrous; stigma 2-branched. Capsules ovoid, 3–4 mm long, 2-valved. Seeds 4–6, oblong, with long, silky, deciduous hairs.

Fls. & Frts.: Oct.–Dec. *Exsiccata*: SUK 1572

Frequent, Riparine plant.

Occurrence: Pavas, Sawantwadi, Kolhapur.

Ecological Note: It grows on stream banks and riverbeds.

TAMARICACEAE

Tamarix ericoides Rottl. in Ges. Naturf. Fr. Berl. Neue. Schr. 4: 214. 1803; T. Dyer in Hook. f. Fl. Brit. India 1: 249. 1874; Shetty and Pandey in Sharma *et al.* (ed.) Fl. India 3: 6. 1993.

Much branched shrubs. Leaves minute, ovate-lanceolate, 1–5 mm long, keeled, acuminate at apex, epuncltate. Flowers 5–6 mm across in axillary and terminal, racemes; bracts ovate, 2–3 mm long, membranous, acuminate. Capsules ovoid, dull, yellowish white, 3-valved. Seeds long with a broad plume, axis feathered with dirty white hairs

Fls. & Frts.: Sept.–Mar. *Exsiccata*: SUK 1487

Rare, riparine plant.

Occurrence: Roha.

Ecological Note: It grows in the riverbeds and water logged fields.

VERBENACEAE

Vitex leucoxylon L.f., Sp. Pl. 293. 1781; Cl. in Hook. f. Fl. Brit. India 4: 587. 1885. *Wallrothia leucoxylon* (L. f.) Roth, Nov. Pl. Sp. 391. 1821.

Large deciduous tree; young branches more or less pubescent. Leaves 3–5-foliolate; leaflets petiolulate, oblong, subobtuse, the terminal leaflet usually the largest, reaching 7.5–10 × 2.5–3 cm, all glabrous above, pale and more or less hairy beneath on midrib, base usually acute, often unequal sided. Flowers fragrant, in axillary, pedunculate lax corymbose cymes; peduncles 2.5–6.5 cm long; bract 2 mm long, lanceolate, caducous. Drupe 2 cm long, succulent, obovoid, blunt, dark purple when ripe, supported on the enlarged flattened calyx; stone large, 4-celled and 4-seeded.

Fls. & Frts.: Feb.–Jul. *Exsiccata*: SUK 2217

Common, riparine plant.

Occurrence: Amboli Tillari.

Ecological Note: It grows in the riverbeds.

Monocots

PANDANACEAE

Pandanus furcatus Roxb. Fl. Ind. 3: 744. 1832. Hook. f. Fl. Brit. India. 6: 484. 1893 *p.p.*; Cook, Aquat. Wetl. Pl. Ind. 1996.

Tree; trunk 3–9 m high, with arial roots. Leaves linear, lanceolate, acute, base caudate–acuminate, margin and midrib armed with short spines. Male flowers in compound cylindric spikes, densely floriferous. Female flowers in solitary spadix; syncarpium ripens orange red. Drupe obconic with convex crown.

Fls. & Frts.: October–November. *Exsiccata*: 544

Common, along streams.

Occurrence: Nerur–Par.

Ecological Note: It grows along fresh water rivers and streams.

Pandanus odoratissimus L. f. Suppl. 64. 1781. *P. fascicularis* Lam. Encycl. 1: 372. t. 1. 1785. KEURA

Shrubby, 6 m high; stem supported by aerial roots. Leaves glaucous–green, 90–150 m long, ensiform, caudate acuminate, marginal spines pointing forward. Male flowers: spadix with numerous, subsessile, cylindric spike enclosed in long, white, fragrant, caudate–acuminate spathes. Female flowers: spadix solitary. Syncarpuim oblong or globose, yellow or red. Drupes numerous, each consisting of 5–12 carpels.

Fls. & Frts.: March–December. *Exsiccata*: 545

Common, along streams.

Occurrence: Pudubudri, Ratnagiri, Udupi.

Ecological Note: It grows along fresh water rivers and streams.

Mangroves

ACANTHACEAE

Acanthus ilicifolia L., Sp. Pl. 639. 1753; Hook. f. Fl. Brit. India 4: 481. 1884. *Dilivaria ilicifolia* Juss. Gen. 103. 1789.

Small shrub. Stem erect, sparsely branched, cylindric, stout, glabrous. Leaves like those of the holly, 7.5–12.5 × 5–6.5 cm., oblong or eliptic, usually pinnatifid or toothed, rigid, acute or truncate at the apex, with sharp spined teeths on the margin terminating the lateral nerves and midrib. Flowers sessile in opposite pairs or terminal crowded spike, varying much in length; bracts 0.5–1.6 cm long. Capsule 2.5 cm long, oblong. apiculate, brown, smooth, shiny. Seeds 1/4.

Fls. & Frts.: Aug.–Mar. *Exsiccata*: 5399

Common, mangroves.

Occurrence: Ratnagiri, Maxem, Panjim, Manglore.

Ecological Note: It grows in estuarine conditions.

APOCYNACEAE

Cerbera manghas L., SP. Pl. 208. 1753. *C. odollam* Gaertn. Fruct. 2: 193. 1791; Hook. f. Fl. Brit. India 3: 638. 1885.

A small tree with acrid, milky juice; branchlets whorled. Leaves alternate, closely set at the ends of branches, coriaceous. Flowers large in terminal peduncled, panicualte cymes. Sepals 5, linear-oblong, very acute, often recurved. Corolla tube long, the upper part dilated, the throat nearly closed by 5 pubescent projecting winglike ribs; lobes white, longer than the tube, broadly elliptic, oblique; tube yellow in throat. Fruit subglobose, green drupe.

Fls. & Frts.: June–Mar.

Rare, mangroves.

Occurrence: Maxem, Manglore, Murideshwar, Udupi.

Ecological Note: It grows in estuarine conditions.

AVICENNIACEAE

Avicennia marina (Forssk.) Vierh. in Denskr. Akad. Wien. Math. Nat. 71: 435, 1907. *Seeura marina* Forsk. Fl. Aegypt., Arab. 37. 1775. *Avicennia alba* auct non Blume, 1825. *A. officinalis* L. var. *alba* Hook.f. in Fl. Brit. India 604. 1885. *A. marina* (Forssk.) Vierh. var. *acutissima* Stapf & Mold. in Phytologia 1: 411. 1940 & 7: 225. 1960.

A large shrub or small tree. Leaves coriaceous, opposite, entire, elliptic-oblong, very obtuse, smooth, shining and glabrous above, thinly tomentose beneath. Flowers numerous, sessile in capitate, pedunculate heads of close cymes, paired in penultimate axils or terminal in trichotomous corymbs. Calyx short, 5-partite; sepals broadly ovate or suborbicular, obtuse, concave, ciliate. Corolla lobes 4, ovate, acute, subequal. Stamens exserted. Capsules ovoid, compressed, beaked, green, wrinkled especially at base.

Fls. & Frts.: April–June. *Exsiccata*: 5465

Frequent, mangroves.

Occurrence: Ratnagiri, Maxem, Mumbai, Panjim, Kudal, Bincholim, Udupi, Manglore.

Ecological Note: It grows in estuarine conditions.

Avicennia officinalis L., Sp. Pl. 110. 1753; Hook. f. in Fl. Brit. India 4: 604. 1885. *A. tomentosa* Jacq. Enum. Pl. Carib. 25. 1760.

A small tree. Leaves coriaceous, opposite, entire, lanceolate, very acute, smooth, shining and glabrous above, whitish tomentose beneath. Flowers numerous, sessile in capitate, pedunculate heads of close cymes, peduncles angular. Calyx short, 5-partite; sepals broadly ovate or suborbicular, obtuse, concave, ciliate. Corolla lobes 4, ovate, acute, subequal. Stamens exserted. Capsules ovoid, compressed, apiculate, bright yellow.

Fls. & Frts.: March–July. *Exsiccata*: 5483

Rare, mangroves.

Occurrence: Ratnagiri, Maxem, Bincholim.

Ecological Note: It grows in estuarine conditions.

CHENOPODIACEAE

Suaeda maritima (L.) Dumort. Fl. Belg. 22, 1827; Hook. f. Fl. Brit. India 5: 14, 1883. *Chenopodium maritimum* L., Sp. Pl. 221. 1753. *Salsola indica* Willd., Sp. Pl. 1317. 1797.

Small annual herbs; stem woody at the base, with numerous very slender branches. Leaves numerous, fleshy, ternate, terete. Flowers minute, axillary, bracteate and bracteolate in small few flowered globose clusters forming very slender elongate, lax spikes. Perianth short, globose, 5-partite. Stamens 5; filaments short. Stigmas 2, long, slender. Seeds obliquely ovoid, suborbicular.

Fls. & Frts.: May–July.

Rare, mangroves.

Occurrence: Alibag.

Ecological Note: Grows in estuarine conditions.

COMBRETACEAE

Lumnitzera racemosa Willd. in Ges. Naturf. Fr. Neue Schr. 4: 187. 1803.

A large shrub or small tree. Leaves sessile, fleshy, crowded towards the end of branches, emarginate at apex, narrowed at base. Flowers many in short axillary spikes. Calyx tube elongate, attenuated at both ends, with two adnate bracteoles at base. Petals 5, white, oblong, subacute. Stamens 10, almost equalling the petals; the alternate filaments shorter. Fruit ovoid, crowned by persistent calyx.

Fls. & Frts.: Jan.–May. *Exsiccata*: 5390

Rare, mangroves.

Occurrence: Manglore.

Ecological Note: Generally grows in estuarine conditions.

EUPHORBIACEAE

Excoecaria agallocha L. Syst. Nat. (ed. 10.), 1288, 1759; Hook. f. Fl. Brit. India 5: 472, 1886.

Tree with white, acrid juice. Leaves alternate, thickly coriaceous, elliptic, acuminate, entire or shallowly sinuate-crenate, base acute. Flowers minute, fragrant, yellowish-green, polygamous in lateral axillary or terminal spikes. Male flowers: sessile in catkin like spikes; bracts with one flower and several minute bracteoles. Sepals 3. Petals absent. Stamens 3, anthers didymous. Female flowers: pedicellate in spikes, upper flowers usually imperfect. Calyx 3-fid. Petals absent. Ovary 3-celled. Fruit a 3-coccus capsule.

Fls. & Frts.: July–August. *Exsiccata*: 5510

Common, mangroves.

Occurrence: Panjim, Manglore.

Ecological Note: It grows in estuarine conditions.

FABACEAE

Derris trifoliata Lour. Fl. Cochinch. 433. 1790; Thothathri, Bull. Bot. Surv. India 3: 185. 1962. *D. uliginosa* Benth. Pl. Jungh. 1: 252. 1852; Hook.f. in Fl. Brit. India 2: 241. 1878. *Pongamia uliginosa* DC. Prodr. 2: 416. 1825.

A large twining shrub. Stem slender, prickly and lenticellate. Leaves imparipinnate. Leaflets 3 pairs, ovate, entire, with undulate margins, rounded at base, acuminate at apex. Flowers numerous in axilary racemes. Calyx campanulate, lobes short. Corolla rose-red. Pod stalked, obliquely rounded, thin, flat, whitish, prominently veined with a distinct narrow wing to the upper suture. Seed usually one, rarely 2, dark brown, flat, reniform.

Fls. & Frts.: February–March. *Exsiccata*: 5607

Common, mangroves.

Occurrence: Ratnagiri, Maxem, Panjim, Manglore.

Ecological Note: It grows in estuarine conditions.

RHIZOPHORACEAE

Aegiceras corniculata (L.) Blanco. Fl. Filip. (ed. 1), 79. 1837. *Rhizophora corniculata* L., Sp. Pl. 635, 1753. *A. majus* Gaertn., Fruct. 1: 216. 1788; Hook. f. Fl. Brit. India 3: 533. 1882.

Large shrub or small tree. Leaves obovate, obtuse at apex, glabrous on both surfaces. Flowers in terminal or leaf-opposed umbels. Flowers sessile, fragrant, greenish-white. Calyx persistent. Petals thick, reflexed. Fruits falcately curved, sharply pointed.

Fls. & Frts.: October–January. *Exsiccata*: 5625

Common, mangroves.

Occurrence: Ratnagiri, Panjim, Manglore.

Ecological Note: It grows in estuarine conditions.

Bruguiera gymnorrhiza Lamk. Tab. Encycl. 2: 517. 1793; Hook. f. Fl. Brit. India 2: 437. 1885.

A glabrous evergreen shrub or tree. Leaves elliptic or elliptic-oblong, acute or slightly acuminate, dark green above, paler beneath. Flowers solitary, axillary, showy. Calyx campanulate, orange-yellow, ribbed. Petals shorter than calyx lobes and as much as sepals. Stamens twice as many as petals and opposite to them. Fruit less than 2.5 cm long, obconic.

Fls. & Frts.: December–April. *Exsiccata*: 5660

Common, mangroves.

Occurrence: Bhatey, Maxem, Pavas, Manglore.

Ecological Note: It grows in estuarine conditions.

Bruguiera parviflora (Roxb.) Wt. & Arn. Prodr. 311. 1834; Hook. f. Fl. Brit. India 2: 438, 1883.

A large shrub. Branches with scars of fallen leaves and stipules. Leaves oblong-lanceolate, acute, shining above, dull beneath, glabrous. Flowers 2–5 in erect cymes longer than petioles. Calyx tube cylindric, fusiform, ribbed; lobes 8. Petals as many as but shorter than the calyx lobes. Stamens twice as many as the petals and opposite to them; filaments of each pair unequal. Fruit concealed by enlarged calyx.

Fls. & Frts.: December–April. *Exsiccata*: 5392

Common, mangroves.

Occurrence: Panjim, Manglore, Malwan.

Ecological Note: Generally grows in estuarine conditions.

Ceriops tagal Robins. in Philipp. J. Sci. 3: 306. 1908.

A large shrub or small tree. Leaves coriaceous, obovate-oblong, rounded at apex, attenuated at the base. Flowers in axillary cymes.

Sepals 5, oblong, subobtuse. Petals 5, white, shorter than sepals. Stamens 10, about as long as the petals, alternately long or short. Fruit slightly conical, gradually thickening to the apex and the acute, deeply grooved and angled.

Fls. & Frts.: July–February. *Exsiccata*: 5480

Common, mangroves.

Occurrence: Ratnagiri, Maxem, Alibag, Manglore.

Ecological Note: It grows in estuarine conditions.

Kandelia candel (L.) Druce. Rep. Bot. Exch. Club. Br. Isl. 1913 (3): 420. 1914; Hook. f. Fl. Brit. India 2: 437. 1885. *Rhizophora candel* L. Sp. Pl. 443. 1753. *Kandelia rheedei* Wt. & Arn. Prodr. 311. 1834.

A small tree. Leaves elliptic-oblong, opposite, obtuse, dark green above, reddish brown beneath. Flowers 4–9, in axillary cymes, usually longer than petioles. Sepals 4–6, surrounded at base by connate bracteoles, lobes narrow, linear, acute. Petals as many as sepals, bifid, lobes divided into numerous, capillary segments. Stamens numerous, of unequal lengths. Fruit clavate.

Fls. & Frts.: March–June. *Exsiccata*: 5481

Common, mangroves.

Occurrence: Ratnagiri, Maxem, Panjim, Manglore.

Ecological Note: It grows in estuarine conditions.

Rhizophora mucronata Poir. in Lamk. Encycl. 6: 189. 1804.

Large evergreen shrub or small tree; young branches marked with scars of fallen leaves and stipules. Leaves elliptic, usually mucronate, glabrous, bright green above, paler and dotted with black beneath. Flowers 3 in axillary cymes slightly longer than petioles. Calyx 4-lobed, lobes pale yellow, coriaceous, triangular, oblong. Petals white, shorter than calyx lobes, oblong, obtuse, thick and fleshy, densely villous on the margins. Stamens 8, anthers linear, acute. Fruit ovoid-conical, surrounded by persistent calyx.

Fls. & Frts.: August–March. *Exsiccata*: 5493

Common, Mangroves.

Occurrence: Ratnagiri, Maxem, Panjim, Manglore.

Ecological Note: Generally grows in estuarine conditions.

SONNERAT IACEAE

Sonneratia apetala Buch.-Ham. in Syme, Emb. Ava 3: 313. 1800; Hook. f. Fl. Brit. India 2: 579. 1879.

A medium sized tree, branches drooping. Leaves opposite, coriaceous, oblong-lanceolate, obtuse, much attenuated at base into short petiole. Flowers solitary or 3 together in axillary or terminal cymes. Calyx lobes 4, coriaceous, campanulate, triangular-ovate. Petals absent. Stamens indefinite, inserted in the throat of calyx tube, filaments reflexed. Berry depressed globose, tipped with style, with thick persistent calyx.

Fls. & Frts.: Feb.–Jun. *Exsiccata*: 5498

Common, mangroves.

Occurrence: Ratnagiri, Bincholim, Guhagar, Manglore, Roha.

Ecological Note: Generally grows in estuarine conditions.

Sonneratia caseolaris (L.) Engl. & Prantl. in Pflanzenr. Nacht. 1: 261. 1897. *S. acida* L.f. Sp. 252. 1781; Hook. f. Fl. Brit. India 2: 579. 1879. *Rhizophora caseolaris* L. in Stickhm, Herb. Amb. 13, 1754. *S. alba* Smith in Rec. Cycl. 33: 2, 1816.

A small tree. Leaves sessile, subfleshy, elliptic-oblong or obovate, obtuse, much attenuated at base. Flowers large, terminal, solitary, on very short, thick pedicle. Calyx campanulate, tube hemispheric, lobes 6, lanceolate, acute. Petals 6, linear-oblong, dark rose coloured. Style very long. Fruit cushion shaped, supported by persistent calyx and tipped by persistent style base in the centre.

Fls. & Frts.: February–August. *Exsiccata*: 5659

Common, mangroves.

Occurrence: Manglore, Panjim, Malwan.

Ecological Note: It grows in estuarine conditions.

REFERENCES

Anonymous, 1992. *Global Biodiversity Strategy*. WRI, IUCN, and UNEP.

Arber, A., 1920. Water plants: A study of aquatic angiosperms. University Press, Cambridge (Reprint 1963. J. Carmer Weinheim Historiae Naturalis classica. Tomus. xxii)

Bruggen, 1968. Revision of Genus *Aponogeton* L. The species of Madagascar. Blumea 16: 243–263.

Biswas and Calder, 1936. Handbook of common water and marsh plant of India and Burma. New Delhi.

Boyd, C.E., 1970c. The vascular aquatic plants for mineral nutrient removal from polluted water. *Eco. Bot.* **24:** 95–103.

Boyd, C.A., 1972. A bibliography of interest in the utilization of aquatic plants. *Eco. Bot.* 26.

Chatrath, K.J.S. (ed.), 1992. *Wetlands of India*. Ashish Publishing House, New Delhi.

Chavan, A.R. and Sabnis, S.D., 1961. Study of Hydrophytes of Baroda and environs. J. Ind. Bot. Soc. **40:** 121–130.

Cook, C.D.K., Gut, B.J., Rix, E.M., Schneller, J. and Seitz, M., 1974. Water plants of world: A manual for the identification of the genera of fresh water macrophytes. W. Junk, The Haque.

Cook, C.D.K., 1979. A revision of the genus Rotala (Lythraceae) Boissiera **29:** 1–156.

Cook, C.D.K. and Ruth Lound, 1982. A revision of genus *Nechamandra* (Hydrocharitaceae). *Aquatic Botany*. **15:** 1–52.

Cook, C.D.K., 1983. A revision of genus *Blyxa* (Hydrocharitaceae). *Aquatic Botany*. **15:** 1–52.

Cook, C.D.K., Symoens, J. and Urmi-Kong, K. 1984. A revision of the genus *ottelia* (Hydrocharitaceae) Aquatic Botany. **18:** 268–274.

Cook, C.D.K., 1996. Aquatic and wetland plants of India. Oxford Univ. Press. Delhi.

Dugen, P.J., 1990. *Wetland Conservation*: A review of current issues and required action GLAND, IUCN.

Gopal, B. (ed.), 1990. Ecology and management of aquatic vegetation in the Indian subcontinent. Academic publisher, Dehradun.

Gupta, R.S., 1965. A study of the hydrophytes and marshy plants of Kota and environs (India) Trop. Ecol. **7:** 153–160.

Joseph, K.T. and Sivarajan, V.V., 1988. *Rotala cookii* a new species of Lythraceae from India showing Hippuris syndrome. Pl. Syst. Evol. **159:** 141–144.

Joseph, K.T. and Sivarajan, V.V., 1990. A new species of *Nymphoides* from India. Nord. J. Bot. 10 (3)

Joseph, K.T., 1991. *Nymphoides sivarajanii,* a new species from India. Wildenowia 20.

Karthikeyan, S., Anand Kumar and Sharma, B.D., 1982. Aquatic angiosperms of Maharashtra (India) J. Econ. Tax. Bot. **3(2):** 423–446

Kaul, V. and Zutshi, D.P., 1967. Study of aquatic and marshland vegetation of Srinagar. Proc. Nat. Inst. Sci. India **33B:** 111–127.

Kaushik, J.P., 1969. The vegetation of marshes, ponds and river bank in Shivpuri, Madyapradesh. Bull. Bot. Surv. India **11:** 84–88.

Little, E.C.S., 1968. Handbook of Utilization of Aquatic Plants. Rome, Food and Agriculture Organisation of the United Nations.

McNabb, C.D., Teirnay, D.P. and Kosek, S.R., 1971. The uptake of Phospharus by *Ceratophyllum demersum* from waste water. Project A–031–Mich. Of the Inst. Wat. Research, Michigan State Univ.

Majumdar, N.C., 1965. Aquatic and Semi-aquatic flora of Calcutta and adjacent localities. Bull. Bot. Soc. Bengal **19:** 10–17.

Manilal, K.S. and Sivarajan, V.V., 1978. A contribution to the study of hydrophytic flora of Kerala. Proc. Natn. Acad. Sci. India **45B:** 225–231.

McNeely, J.A., Miller, K.R., Reid, W., Mittermeiror, R. and Warmer, T., 1990. *Conservation of the World Biodiversity*. Washington, DC. IUCN. World Resources Institute, World Bank. WWF. US and Conservation International.

McNeely, J.A., 1992. The sinking arc: pollution and the worldwide loss of biodiversity. Biodiversity and Conservation **1:** 2–28.

Mitchell, D.S., 1974. Aquatic vegetation and its use and controll. UNESCO. Paris.

Mohan Ram, H.Y., 1978. In vitro culture of Aquatic Flowering Plants: Achievements and prospets. 10th P.R. White Memorial Lecture. Ahmedabad, India. 1–17.

Naskar, K.R., 1990. *Aquatic and semiaquatic plants of the lower Ganga Delta.* Daya Publishing House, Delhi.

Raghvan, R.S. Kulkarni, A.R. and Yadav, S.R., 1982. *Aponogeton* satarensis A new species from India. Kew. Bull. **36:** 687–689

Rai, U.N., Sinha, S., Tripati, R.D. and Chandra, P., 1995. Waste water treatability potential of some macrophytes: Removal of heavy metals. Ecological Engineering **5:** 5–12.

Rao, R.R. and Sharma, B.D., 1990. *A manual for herbarium collections.* BSI. India.

Sahai, R. and Sinha, A.B., 1968. A supplement to the aquatic and swampy vegetation of Gorakhpur. *Indian Forester* **9(4):** 819–821.

Satyanarayana, G. 1962. Hydrophyte vegetation of Jhalukbari. *Bull. Bot. Surv. India* **4:** 217–218.

Sculthorpe, C.D., 1967. *The biology of aquatic vascular plants* Edward Arnold, London.

Seidel, K., 1971. Macropyhtes as functional elements in the environment of man. *Hydrobiology* **12:** 121–30.

Seerwani, A.B., 1962. Study of hydrophytes and plants of low lying habitats in Jabalpur. Ibid. **4:** 227–231.

Sivrajan, V.V. and Joseph, K.T., 1993. The genus *Nymphoides* Seguier in India. *Aquatic Botany;* **45:** 145–170.

Stafleu, F.A., 1978. International code of Botanical Nomenclature adapted by the 12th International Botanical Congress, Leningard, July 1975 IAPT. Utrecht.

Subramanyam, K., 1962. *Aquatic Angiosperms.* C.S.I.R. New Delhi.

Vyas, L.N., 1964. A study of the hydrophytes and marsh plants of Alwar and environs. *J. Ind. Bot. Soc.* **43:** 17–30.

Yadav, S.R. and Govekar, R.S., 1994. *Aponogeton bruggenii,* A new species from India. *Rheedea* **4:** 17–20.

Yount J.L. and Crossaman, R.A., 1970. Eutrophication control by plant harvesting. *J. Wat. Pollt. Cont.* **42:** 173–83.

INDEX TO SPECIES AND FAMILIES

Italicized Names Indicate Synonyms